T A K E
A
L O O K
A T
A
G O O D
B O O K

The third collection of additive alphametics for the connoisseur

STEVEN KAHAN

Library of Congress Catalog Card Number: 96-16547
ISBN: 0-89503-142-6 (pbk.)

Library of Congress Cataloging-in-Publication Data

Kahan, Steven.
Take a look at a good book : the third collection of additive alphametics for the connoisseur / by Steven Kahan.
p. cm.
ISBN 0-89503-142-6 (pbk.)
1. Mathematical recreations. I. Title.
QA95.K324 1996
793.7'4- -dc20 96-16547
CIP

acknowledgments

Certain individuals and groups of individuals deserve special kudos. I am sincerely grateful to

Stuart Cohen, President of Baywood Publishing, who is truly a man of his word.

Bobbi Olszewski of Edco, whose meticulous attention to detail never ceases to amaze me.

Dover Publications, for allowing me to include several excellent anagrams chosen from Howard Bergerson's *Palindromes and Anagrams*, published in 1973.

my wonderful and loving family, for affording me the time and opportunity to create and solve problems other than those that occur in everyday life.

S.K. 1996

dedication

directed approach: page 79
solution: page 113

A E G L R S T U

0 1 2 3 4 5 6 7 8 9

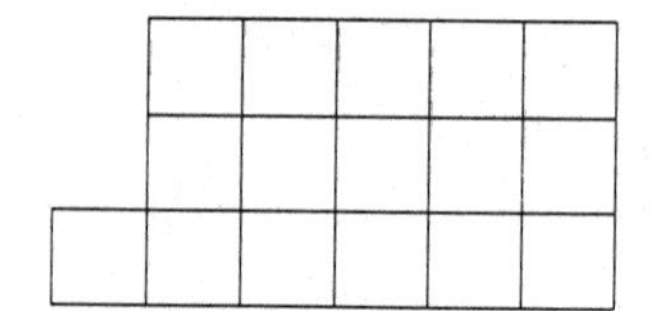

dedication

Leonhard Euler (1707-1783) is generally acknowledged to have systematized mathematics as we know it today. His built-in mental computer enabled him to perform complicated calculations without the aid of pencil and paper. It has been said that Euler wrote mathematics as effortlessly as most men breathe.

Johann Friedrich Carl Gauss (1777-1855), the Mozart of mathematics, was a true child prodigy with an amazingly accurate memory. His "Disquisitiones Arithmeticae," published in 1801, has been widely acclaimed as the most important number theory book ever written.

We are indebted to these two giants for their insights, inspirations, and immeasurable contributions to our present body of knowledge in mathematics. Accordingly, this book is proudly dedicated to

and

$$\begin{array}{r} \text{E U L E R} \\ \underline{\text{G A U S S}} \\ \text{G R E A T S} \end{array},$$

in the mathematical world. It is certainly appropriate that these GREATS should be greatest in the desired solution.

Dedication

table of contents

preface 1

narrative alphametics

- greetings 7
- threesomes 11
- from A to Z 17
- pun, pun, pun 21
- on a count 23
- array, array 25
- dynamic duos 31
- marking time 35
- in a word 39
- fifty's not nifty 41
- as a matter of fact 43
- rhyme time 45
- identity crisis 49
- generation gap 51
- all mixed up 55
- fourplay 59
- whatchamacallits 65
- home sweet home 67
- ready, set, go 71
- nothing but the truth 75

ideal doubly-true alphametics

- i.d.t. – fiftytwo 9
- i.d.t. – eightysix 9
- i.d.t. – fourhundredfortyone 15
- i.d.t. – sixtynine 17
- i.d.t. – eighty 19
- i.d.t. – seventy 21
- i.d.t. – ninetyone 29
- i.d.t. – ninetyfive 29

i.d.t. – thousand 33
i.d.t. – ninety 35
i.d.t. – eightynine 37
i.d.t. – hundred 43
i.d.t. – ninetyeight 47
i.d.t. – ninetynine 57
i.d.t. – sixtysix 61
i.d.t. – ninetytwo 63
i.d.t. – ninetysix 69
i.d.t. – eightytwo 73

solutions chart 119

about the author 121

tantalizing numerical tidbits (TNTs) can be found on the following pages:

10, 12, 19, 24, 27, 30, 33, 38, 40, 41,
47, 50, 52, 54, 58, 59, 61, 63, 64, 69.

preface

It isn't easy to write introductory comments for the third book in a series. Anything that needs to be said has already been said once or twice before, and readers who possess the previous volumes are not at all interested in still another rehashing of old material. Nonetheless, a book without some sort of preface is like a building without a foundation. In both instances, structural integrity is dependent upon a solid underpinning. If you consider yourself to be a grizzled alphametics veteran, skim or skip what follows here and get right down to business. Neophytes, however, are urged to read on without interruption. The pointers gleaned might prove to be beneficial when the actual puzzles are ultimately confronted.

What exactly *is* an alphametic? Simply stated, it is a decoding problem requiring the solver to replace letters of the alphabet with appropriately chosen digits so that the resulting numerical example has arithmetic validity. In essence, there are but three rules that govern the entire solution process:

1. once a digit is assigned to a letter, that digit must be assigned to every appearance of that letter in the puzzle;
2. once a digit is assigned to a letter, that digit cannot be assigned to any other letter in the puzzle; and
3. the digit zero cannot be assigned to a letter that appears at the beginning of any word in the puzzle.

An item sent to my attention a year or two ago serves as a fine illustration of the form. Remember the serpent's words to Eve in the Garden of Eden? He implored her to

```
        E A T
  +   T H A T
    ---------
    A P P L E !!
```

Let us investigate the logical process that converts this edict into a conventional addition example.

Since every four-digit number is less than 10,000 and every three-digit number is less than 1,000, the sum of two such numbers is necessarily less than 11,000. This sum, though, is a five-digit number, hence is greater than 10,000. Consequently, A must be 1 and P must be 0. Further, we can conclude that T = 9. Otherwise, we would be adding a number less than 1,000 to one less than 9,000, leaving us short of the requisite total. The units column then produces E = 8 while generating a carryover of 1 into the tens column. Together with the previously-found value of A, we learn from the tens column that L = 3. Finally, the hundreds column yields the equation E + H = P + 10, where the "10" is required to accommodate the needed carryover into the thousands column. When the values of E and P are substituted into this relationship, we get 8 + H = 10, from which it follows that H = 2. Therefore, the unique solution of the puzzle turns out to be

$$\begin{array}{r} 8\ 1\ 9 \\ +\ \ 9\ 2\ 1\ 9 \\ \hline 1\ 0\ 0\ 3\ 8 \end{array} .$$

Needless to say, not all alphametics are this short and sweet. It is far more likely that considerably more gray matter will be expended before a solution in fact emerges.

The popular features of the two earlier books, *Have Some Sums to Solve* and *At Last!! Encoded Totals, Second Addition*, have been retained here. The thirty-eight puzzles presented in Section 1, along with the cover, dedication, and preface puzzles, all fall into the special subcategory of additive alphametics. Each of their sums has a unique decoding, sometimes insured by the imposition of an initial condition. The presence of a pair of interchangeable digits within the summands themselves does not constitute a contradiction of this uniqueness.

Within the subcategory, two varieties of alphametics are included—the ideal, doubly-true type (abbreviated "i.d.t.") and the narrative type. In the former, all ten digits appear in the solution and a mathematically correct addition example results when the problem is read aloud. The latter type is stated within the context of a brain-teaser (which requires solution itself) and/or some informative paragraphs. Blank grids have been supplied so that solutions can be proudly preserved for posterity. Throughout this first section will also be encountered a collection of TNTs, tantalizing numerical tidbits that are guaranteed to raise the eyebrows of integer aficionados. Moreover, these serve to fill those uninteresting spaces that inevitably crop up in page layouts.

Directed approaches to each of the puzzles are offered in Section 2. These strategic outlines enable the solver to skirt around pitfalls and obstacles without removing the challenge associated with the quest for the actual answer. In Section 3, solutions to all puzzles are given. The order of presentation of these solutions is intentionally different from that of the puzzles. This is meant to discourage the temptation of becoming a digital voyeur (an affliction having nothing whatsoever to do with peeking at another's fingers and toes). This section also contains responses to all queries raised within the context of the narrative alphametics. Lastly, a chart is provided that indicates the total number of solutions to each puzzle in the absence of any constraint.

The major difference between this and the two earlier works is the prevalence here of "wider" i.d.t.s, wherein the sums span nine or more columns. With new advances in computer technology and the advent of programs that actually generate solvable i.d.t.s, new totals are being explored and established at a record pace. However, I've scrupulously refrained from using such tools in generating and solving the puzzles in this book. Just call me old-fashioned!

I'll sign off now in what seems to be a fitting manner, wishing you an exciting excursion through this realm of puzzledom called alphametics. Enjoy the adventure!

```
Y O U R S
    V E R Y
T R U L Y ,
```

Steven Kahan

(In the salutation, TRULY is truly odd, even in base 9.)

preface
directed approach: page 80
solution: page 110

E	L	O	R	S	T	U	V	Y
0	1	2	3	4	5	6	7	8

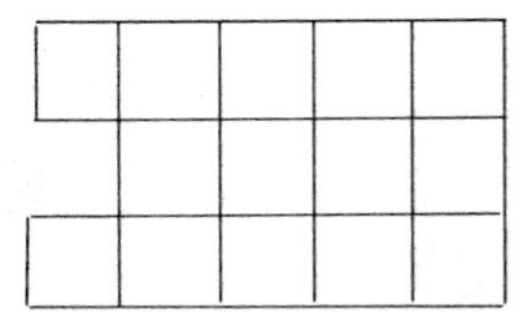

? ? ? ? ? ? ? ? ? ? ? ? ? ? ? ? ? ? ? ?

SECTION 1

PUZZLES

? ? ? ? ? ? ? ? ? ? ? ? ? ? ? ? ? ? ? ?

greetings
directed approach: page 80
solution: page 115

A D E H K L N S

0 1 2 3 5 6 7 8 9

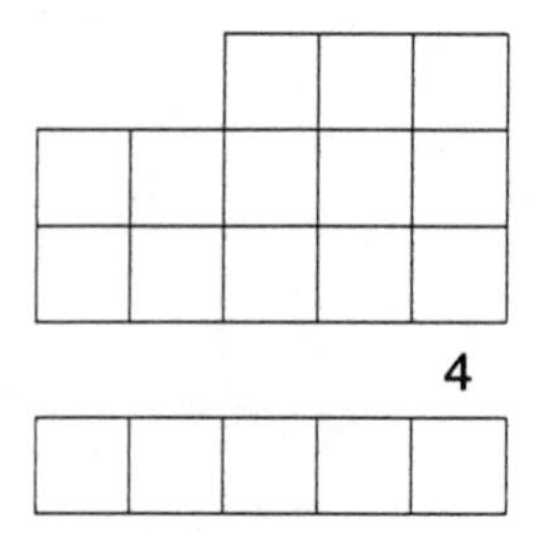

greetings

Master Detective Chuck Chan recently received the following note from a client who delighted in testing the sleuth's deductive mettle:

> Dear Detective Chan:
>
> You and your lovely wife are cordially invited to attend an intimate dinner party at my home with four other married couples. Some of the invitees shall already be acquainted. Those who are not shall be introduced and be asked to exchange handshakes. After the completion of these formalities, each of my ten guests shall be asked to record on an otherwise blank slip of paper the number of handshakes in which he or she partook. When the nine others hand their slips to you, you'll notice that no two of them display the same digit. Given that no individual shall shake his or her hand, nor that of his or her spouse, how many hands shall Mrs. Chan shake? R.S.V.P.!!

After a few contemplative moments, Chuck chuckled aloud, took pen in hand, and filled out the enclosed response card thusly:

> Mrs. Chan and I shall be delighted to attend your party, and

```
      S H E
  S H A L L
  S H A K E
          4
  ---------
  H A N D S .
```

Solve the alphametic without re-employing the digit "4," and then supply the logic that correctly led the inscrutable Chan to his conclusion.

i.d.t. – fiftytwo

directed approach: page 81
solution: page 116

E	F	H	I	N	O	R	T	W	Y
0	1	2	3	4	5	6	7	8	9

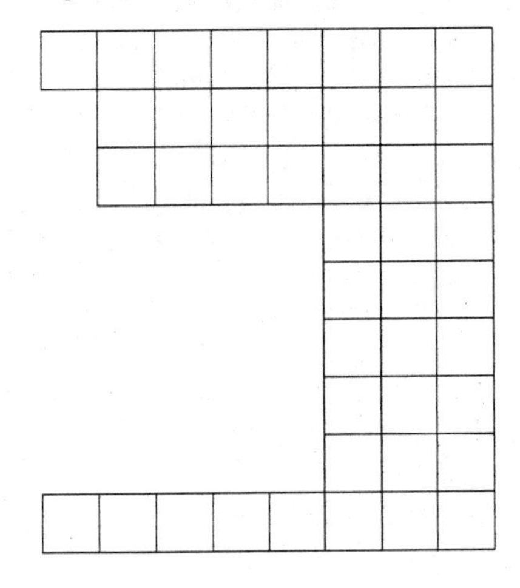

i.d.t. – eightysix

directed approach:
page 81
solution: page 111

E	G	H	I	N	S	T	V	X	Y
0	1	2	3	4	5	6	7	8	9

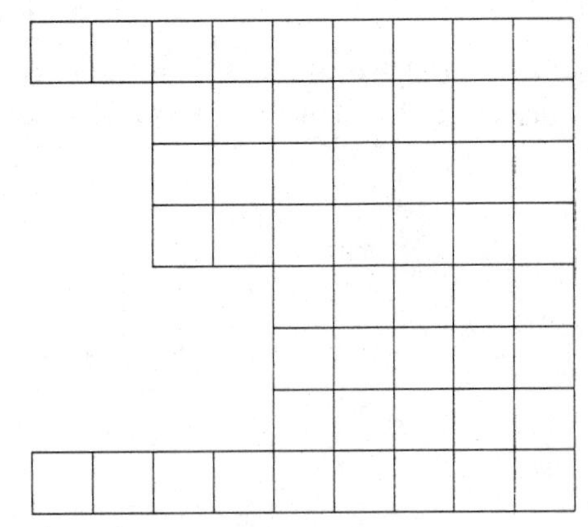

i.d.t. – fiftytwo

THIRTEEN
FIFTEEN
FIFTEEN
TWO
TWO
TWO
TWO
ONE
FIFTYTWO

i.d.t. – eightysix

SEVENTEEN
SIXTEEN
SIXTEEN
SIXTEEN
SEVEN
SEVEN
SEVEN
EIGHTYSIX

A pair of primes that differ by two, such as 29 and 31, are called twin primes. Although it is well known that infinitely many primes exist, it is an open question whether or not there exist infinitely many twin primes. One interesting example is one trillion sixty-one and one trillion sixty-three. It is easy to show that 3, 5, and 7 are the only triplet primes that exist.

threesomes

Perhaps the best-known geometric result ever established is one that dates back to antiquity. More than 2,500 years ago, the Greek mathematician Pythagoras verified that if triangle ABC has a right angle at C, then $a^2 + b^2 = c^2$, where a, b, and c are the lengths of the sides of the triangle that are opposite vertices A, B, and C, respectively. This statement, called the Pythagorean Theorem in honor of its discoverer, has been reconfirmed in many different ways throughout history. One particularly nice proof is a direct consequence of the following sketch:

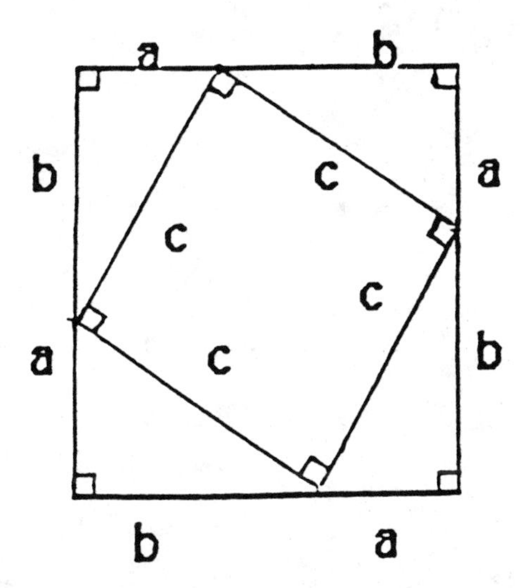

Observing that the area of the large square must equal the sum of the areas of the small square and the four right triangles, we find that $(a + b)^2 = c^2 + 4(\frac{1}{2}ab)$. Upon simplification, this yields the desired result.

A set of positive integers that satisfies the Pythagorean Theorem is called a Pythagorean triplet, with 3, 4, 5 being the most familiar representative of the genre. Once a Pythagorean triplet is known, it spawns infinitely many others. Stated formally, if a, b, c is a Pythagorean triplet, then so is ka, kb, kc for any integer $k \geq 2$. (To see this, simply note that $(ka)^2 + (kb)^2 = k^2a^2 + k^2b^2 = k^2(a^2 + b^2) = k^2c^2 = (kc)^2$.) Thus, 6, 8, 10 is also a Pythagorean triplet, as is 72, 96, 120.

The smallest solution of $a^5 + b^5 + c^5 + d^5 + e^5 = f^5$ in which a, b, c, d, e, and f are distinct positive integers is a = 19, b = 43, c = 46, d = 47, e = 67, f = 72.

The product of a two-digit number and its reversal is never a perfect square, unless the digits are equal. This is not the case for numbers with more than two digits, as $169 \times 961 = 403^2$ and $1089 \times 9801 = 3267^2$.

A primitive Pythagorean triplet, or PPT for short, is one whose members have no common factor other than 1. Interestingly, we can produce PPTs quite easily. Specifically, let m and n be two positive integers such that (i) $m > n$; (ii) m and n have opposite parity; and (iii) m and n have no common factor other than 1. Define

$$a = m^2 - n^2,\ b = 2mn, \text{ and } c = m^2 + n^2.$$

Then a, b, c is always a PPT. To illustrate, we can select $m = 2$ and $n = 1$ and get $a = 3$, $b = 4$, and $c = 5$, the previously-mentioned PPT. Ten others appear in the table below.

m	n	a	b	c
3	2	5	12	13
4	1	15	8	17
4	3	7	24	25
5	2	21	20	29
5	4	9	40	41
6	1	35	12	37
6	5	11	60	61
7	2	45	28	53
7	4	33	56	65
7	6	13	84	85

The revelation of the formulas that generate a, b, and c makes it obvious that the list of PPTs is endless. For instance, if we choose $m = 101$ and $n = 66$, we find the PPT 5845, 13332, 14557. That is, $(5845)^2 + (13332)^2 = (14557)^2$. Check it on your calculator if you're a skeptic by nature. Once you're convinced, you can create PPTs made up of extraordinarily large integers.

threesomes
directed approach: page 82
solution: page 118

A	E	G	I	L	N	P	R	S	T
0	1	2	3	4	5	6	7	8	9

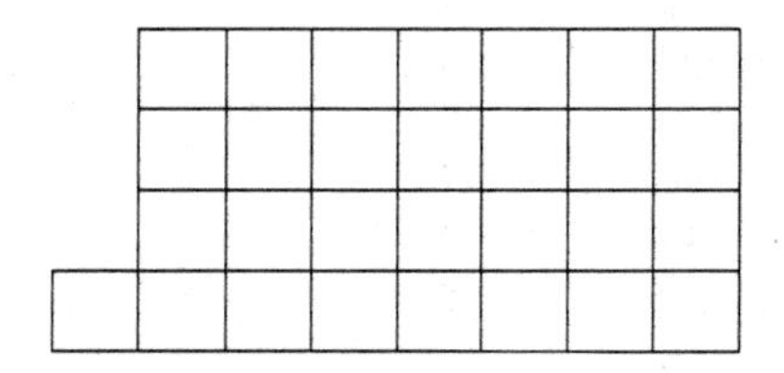

i.d.t. – fourhundredfortyone
directed approach: page 82
solution: page 113

D	E	F	H	N	O	R	T	U	Y
0	1	2	3	4	5	6	7	8	9

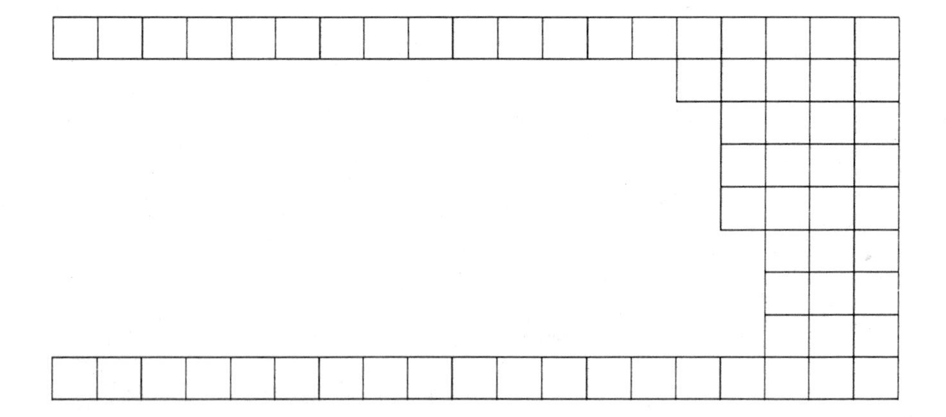

To close this little vignette, here's a question concerning not Pythagorean, but biological triplets. How is it possible for a woman to give birth to three children in a span of five minutes so that no two of the babies have the same birthday?

Such newborns would surely be classified as

```
  S T R A N G E
  S T R A N G E
  S T R A N G E
---------------
T R I P L E T S .
```

Accordingly, odd TRIPLETS are in order here.

i.d.t. – fourhundredfortyone

```
F O U R H U N D R E D F O U R T E E N
                              T H R E E
                                F O U R
                                F O U R
                                F O U R
                                  T E N
                                  O N E
                                  O N E
-----------------------------------------
F O U R H U N D R E D F O R T Y O N E
```

from a to z

directed approach: page 83
solution: page 111

C	H	I	L	M	N	O	R	U	W
0	1	2	3	4	5	6	7	8	9

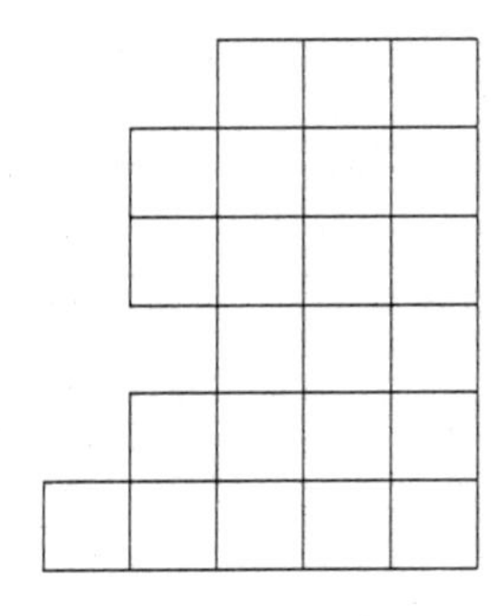

i.d.t. – sixtynine

directed approach: page 83
solution: page 109

E	H	I	N	R	S	T	V	X	Y
0	1	2	3	4	5	6	7	8	9

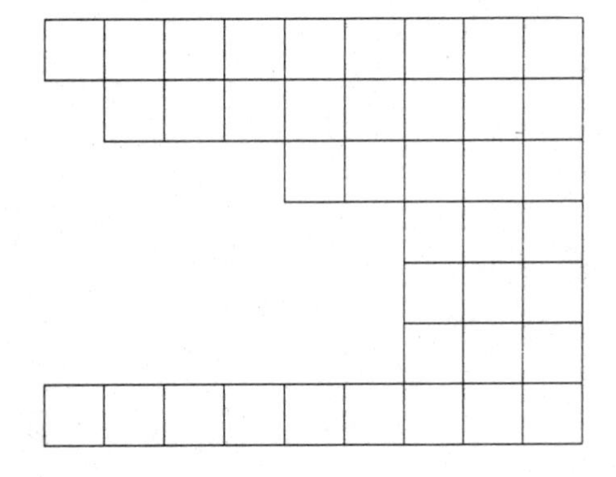

from a to z

A ten-volume set of encyclopedias is displayed on a bookshelf in consecutive order, with Volume 1 at the viewer's left and with no gaps between volumes. The front cover of each of the ten volumes is $\frac{1}{9}$" thick, as is the back cover of each. The thickness of the pages of each of the books is precisely $3\frac{1}{2}$".

An inchworm who also happens to be a voracious bookworm starts on the first page of Volume 1 and proceeds on a horizontal path, systematically munching his way through to the last page of Volume 10. Keeping in mind that the route described suggests that the smallest MUNCH is appropriate,

```
    H O W
  M U C H
  W I L L
    O U R
  W O R M
  -------
M U N C H ?
```

i.d.t. – sixtynine

```
S E V E N T E E N
  N I N E T E E N
        T H R E E
            T E N
            T E N
            T E N
-----------------
S I X T Y N I N E
```

i.d.t. – eighty

directed approach: page 84
solution: page 115

E	F	G	H	I	N	O	T	V	Y
0	1	2	3	4	5	6	7	8	9

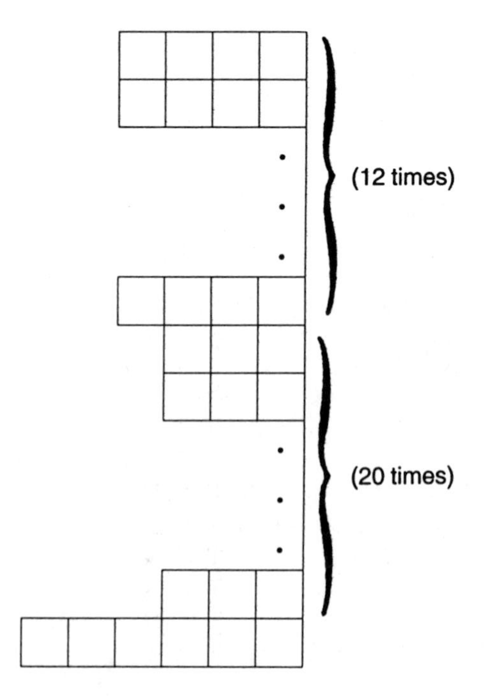

i.d.t. – eighty

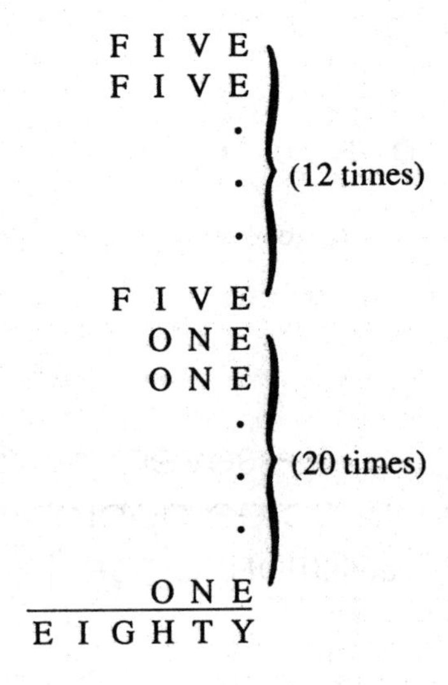

$$2^2 + 6^2 + 22^2 + 262^2 = 2 \times 6 \times 22 \times 262$$

pun, pun, pun
directed approach: page 84
solution: page 116

A	E	H	I	M	N	O	R	S	T
0	1	2	3	4	5	6	7	8	9

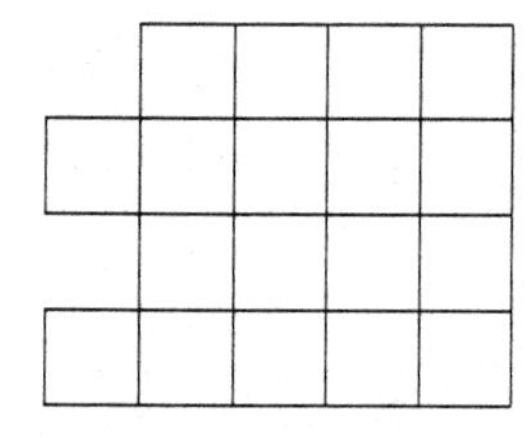

i.d.t. – seventy
directed approach: page 85
solution: page 114

E	H	L	N	R	S	T	V	X	Y
0	1	2	3	4	5	6	7	8	9

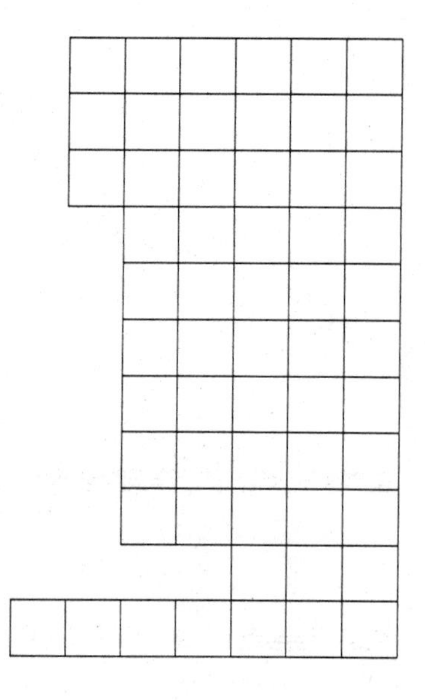

pun, pun, pun

Three brothers purchase a cattle ranch but can't seem to agree on a name for it. Perplexed, they write to their father, an inveterate punster, asking him for a suggestion.

"There is really only one name you should even consider for your new ranch," their dad replied.

Since his

```
  S O N S
R A I S E
  M E A T
---------
T H E R E ,
```

what was the proposal of the paterfamilias? (Like most fathers, this one assumes that his SONS are the greatest.)

i.d.t. – seventy

```
  T W E N T Y
  E L E V E N
  E L E V E N
    T H R E E
    T H R E E
    T H R E E
    T H R E E
    T H R E E
    T H R E E
        T E N
-------------
S E V E N T Y
```

on a count
directed approach: page 85
solution: page 109

A D E H L N O S T

0 1 2 3 4 5 6 7 8 9

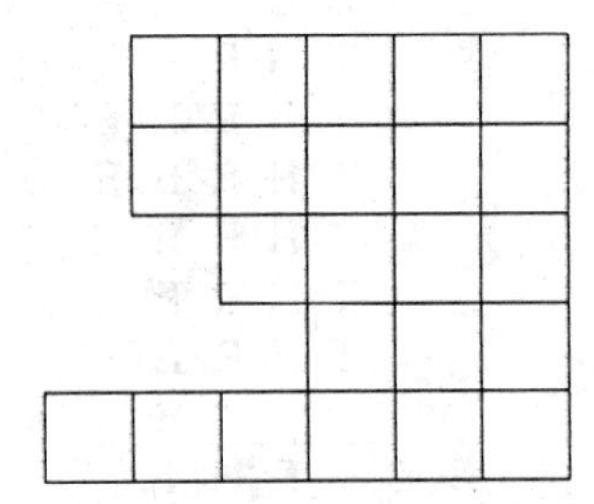

on a count

To reproduce the accompanying figure, subdivide each side of an equilateral triangle into eight equal parts and employ line segments parallel to the sides of the triangle to connect the partition points. Triangles galore are thus created–some, called dels, point to the south, while the rest, called deltas, point northward.

Including the original, how many different triangles are visible? Put another way, your challenge is to

```
 T O T A L
 T H E S E
   D E L S
     A N D
-----------
D E L T A S .
```

Here's a hint—DELS and DELTAS are even, both by actual count and alphametically.

Bonus Question for the Mathematically Muscular:

Had the sides of the equilateral triangle been subdivided into n equal parts and the same procedures were then followed, find a formula that enumerates the total number of triangles in the resulting figure in terms of n.

11826 and 30384 are the smallest and largest integers whose squares display each digit from 1 to 9 exactly once: $(11826)^2 = 139854276$ and $(30384)^2 = 923187456$. There are thirty such integers in all. Similarly, 32043 and 99066 are the smallest and largest integers whose squares display the digits from 0 to 9 once and only once, where 0 is not the leading digit: $(32043)^2 = 1026753849$ and $(99066)^2 = 9814072356$. There are eighty-seven such integers in all.

array, array

The noted mathematical detective Vic Tracy was presented with the following dilemma by his arch-rival, The Baffler:

Consider the 7×7 matrix

$$\begin{matrix} 1 & 8 & 15 & 22 & 29 & 36 & 43 \\ 2 & 9 & 16 & 23 & 30 & 37 & 44 \\ 3 & 10 & 17 & 24 & 31 & 38 & 45 \\ 4 & 11 & 18 & 25 & 32 & 39 & 46 \\ 5 & 12 & 19 & 26 & 33 & 40 & 47 \\ 6 & 13 & 20 & 27 & 34 & 41 & 48 \\ 7 & 14 & 21 & 28 & 35 & 42 & 49 \end{matrix} .$$

I challenge you to select six numbers from this array that simultaneously satisfy three conditions:

1. no two numbers come from the same row;
2. no two numbers come from the same column;
3. in the set of numbers chosen, each of the ten digits appears once and only once.

"A simple matter," announced Vic Tracy without the slightest hesitation. "The numbers to be selected are 7, 9, 18, 26, 30, and 45."

"Not so fast, Tracy," The Baffler replied. "You've chosen 9 and 30, both of which reside in the second row. Perhaps you'd like to reconsider?"

"How careless of me," Tracy retorted. "However, the blunder is easily repaired by just removing 26 and 30 from the set and replacing them with 20 and 36. That should put the problem to bed quite nicely."

"On the contrary, the problem remains wide awake," crowed The Baffler. "Your latest offer contains 18 and 20, two numbers that sit prominently in the third column."

array, array

directed approach: page 86
solution: page 116

C I F K P S V X

0 1 2 3 4 5 6 7 8 9

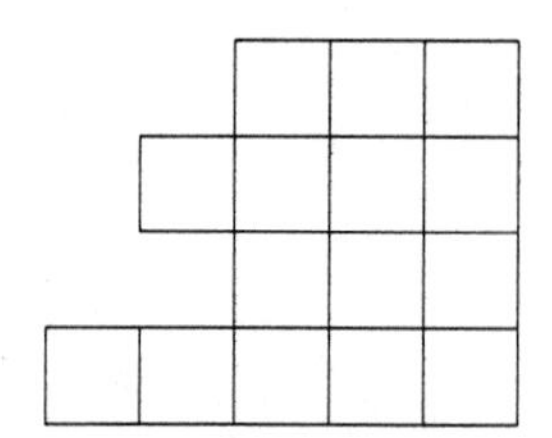

"So they do, so they do," a disappointed Vic admitted reluctantly. After a moment's reflection, he broke into a wide grin and exclaimed, "Eureka! All that needs to be done is to substitute 34 for 30 in my original proposal. The issue is hereby resolved."

"Certainly, no two numbers now come from the same row or the same column," The Baffler said. "However, this new set of numbers contains the digit '4' twice and the digit '0' not at all. The true solution eludes you still."

A crimson-faced Tracy bowed his head and sullenly walked away, muttering that the problem, no doubt, is unsolvable. Wrong again!! There is a solution and a unique one at that. Now it's your chance to be a gumshoe and see if you can

```
      F I X
    V I C' S
      S I X
  -----------
  P I C K S .
```

Here, we request odd PICKS in the decoding, although not in the resolution of The Baffler's challenge.

$$\frac{3^2+4^2+5^2+6^2+7^2+8^2+9^2}{1^2+2^2+3^2+4^2+5^2+6^2+7^2} = 2$$

i.d.t. – ninetyone
directed approach: page 86
solution: page 113

E	F	I	N	O	R	T	U	W	Y
0	1	2	3	4	5	6	7	8	9

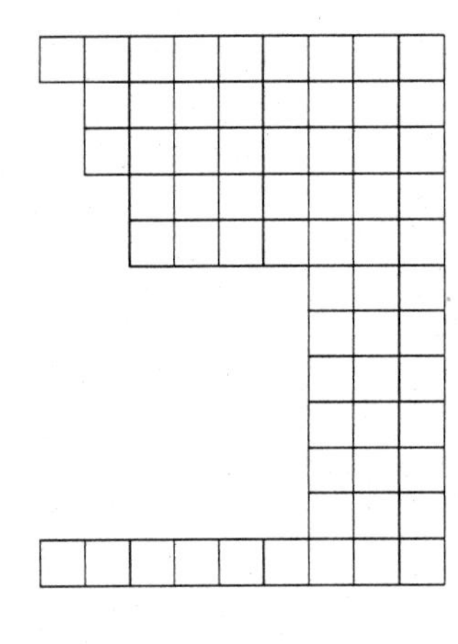

i.d.t. – ninetyfive
directed approach: page 87
solution: page 118

E	F	I	N	S	T	V	W	X	Y
0	1	2	3	4	5	6	7	8	9

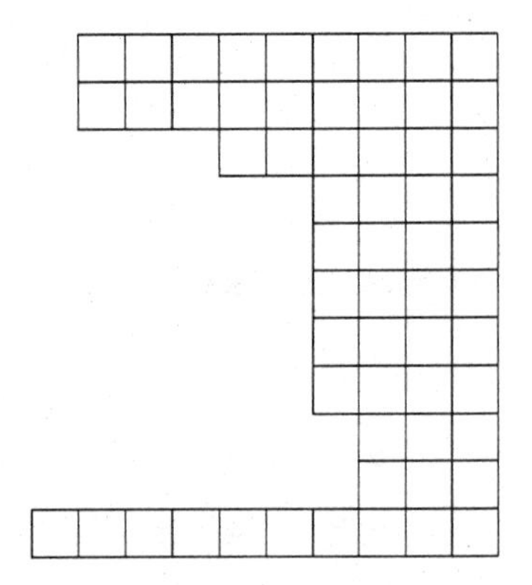

i.d.t. – ninetyone

```
T W E N T Y T W O
  F O U R T E E N
  F O U R T E E N
    F I F T E E N
    F I F T E E N
            T W O
            T W O
            T W O
            T W O
            T W O
            O N E
-----------------
N I N E T Y O N E
```

i.d.t. – ninetyfive

```
  S E V E N T E E N
  S E V E N T E E N
        T W E N T Y
            N I N E
            F I V E
            F I V E
            F I V E
            F I V E
              S I X
              S I X
-------------------
N I N E T Y F I V E
```

dynamic duos
directed approach: page 87
solution: page 110

A	E	H	L	O	R	S	T	V	W
0	1	2	3	4	5	6	7	8	9

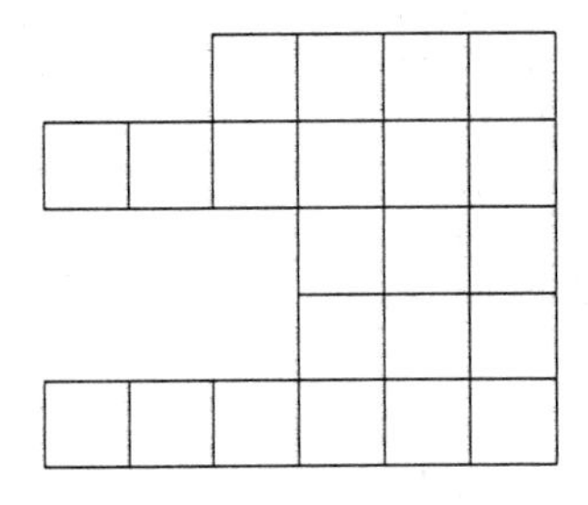

Subfactorial n, denoted by !n, is given by

$$!n = n!\left[1 - \frac{1}{1!} + \frac{1}{2!} - \frac{1}{3!} + \ldots + (-1)^n\left(\frac{1}{n!}\right)\right],$$

where n is a positive integer.

(For example, !1 = 0, !2 = 1, !3 = 2, and !4 = 9.)

Then 148349 = !1 + !4 + !8 + !3 + !4 + !9 is the only integer that is equal to the sum of the subfactorials of its digits.

dynamic duos

In case you don't have any skeletons in your closet, here are some for you to store there. All the consonants have been stripped away from twenty commonplace pairs.

W E' L L
R E V E A L
A L L
T H E
V O W E L S ,

where ALL should be made most of all. You do the rest. A score of 17 or more is _E_ _I_I_!!

1. Literary lovers
   _O_EO and _U_IE_
2. Comedy giants
   _AU_E_ and _A_ _ _
3. Iberian neighbors
   _ _AI_ and _O_ _U_A_
4. Pinafore creators
   _I_ _E_ _ and _U_ _I_A_
5. Shaker fillers
   _A_ _ and _E_ _E_
6. Frequent intro
   _A_IE_ and _E_ _ _E_E_
7. Vacation aim
   _E_ _ and _E_A_A_IO_
8. Market animals
   _U_ _ _ and _EA_ _
9. Frank's mates
   _U_ _A_ _ and _AUE_ _ _AU_
10. Betsy's handiwork
    _ _A_ _ and _ _ _I_E_
11. Sure things
    _EA_ _ and _A_E_
12. Bronte book
    _ _I_E and _ _E_U_I_E
13. Storm warnings
    _ _U_ _E_ and _I_ _ _ _I_ _
14. American explorers
    _E_I_ and _ _A_ _
15. Tingling sensation
    _I_ _ and _EE_ _E_
16. Legally separated
    _ _U_ _ _ and _ _A_E
17. Italian fare
    _EA_ _A_ _ _ and _ _A_ _E_ _I
18. Melodramatic intrigue
    _ _OA_ and _A_ _E_
19. Bone breakers
    _ _I_ _ _ and _ _O_E_
20. Identical twins
    _ _EE_ _E_EE and _ _EE_ _E_U_

i.d.t. – thousand

directed approach: page 88
solution: page 118

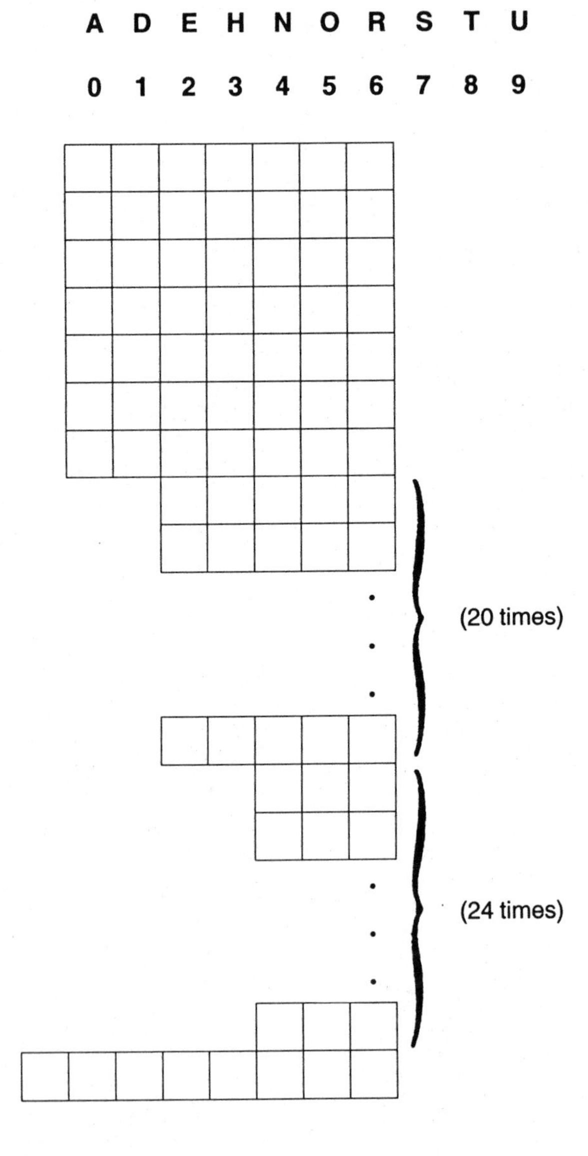

i.d.t. – thousand

```
  H U N D R E D
  H U N D R E D
  H U N D R E D
  H U N D R E D
  H U N D R E D
  H U N D R E D
  H U N D R E D
      T H R E E  \
      T H R E E   |
              .   |
              .   } (20 times)
              .   |
      T H R E E  /
          T E N  \
          T E N   |
              .   |
              .   } (24 times)
              .   |
          T E N  /
---------------
T H O U S A N D
```

$$(6048 + 1729)^2 = 60481729$$

and

$$(5288 + 1984)^2 = 52881984$$

marking time

directed approach: page 89
solution: page 115

A	D	E	G	I	L	R	S	T	W
0	1	2	3	4	5	6	7	8	9

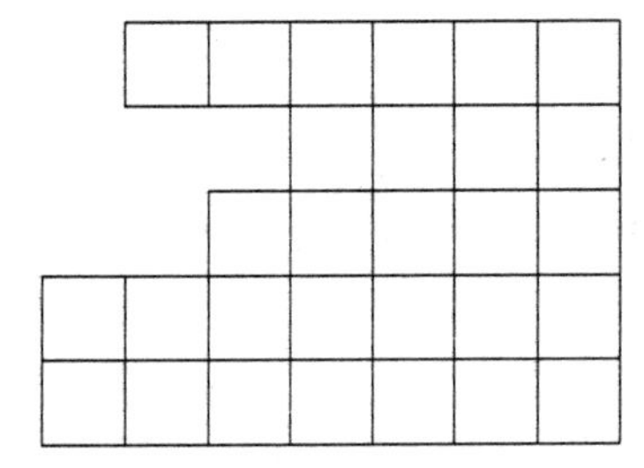

i.d.t. – ninety

directed approach: page 89
solution: page 116

E	F	I	L	N	S	T	V	W	Y
0	1	2	3	4	5	6	7	8	9

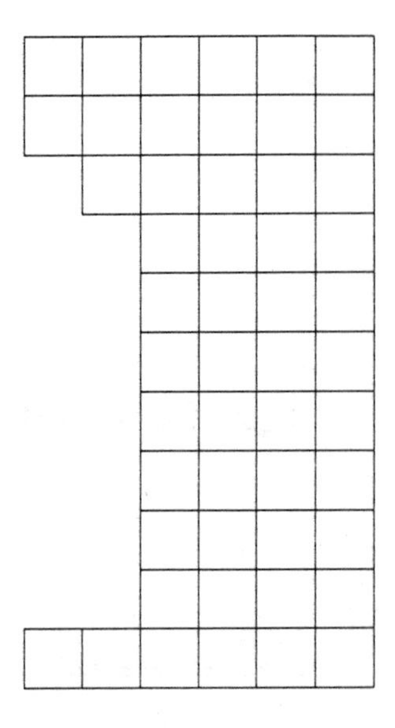

marking time

"It's a very special time of the day," Walter remarked to his wife, Gidget.

"What's so special about it?" Gidget queried.

"Look at the display on your digital clock," instructed Walter. "I'll show you something that I discovered when I was in the army."

Gidget dutifully heeded her husband's request, noting that the clock showed three digits, with the leftmost space vacant.

"I'll now place a certain digit to the far right of the three that you see, thus shifting the entire display one space to the left," continued Walter. "Remarkably, this maneuver will not change the actual time one iota."

As Gidget watched, he proceeded to do just that!

Can you figure out how

```
  W A L T E R
      W I L L
    A L T E R
G I D G E T' S
--------------
D I G I T A L ?
```

If you can, then you'll know the exact time of the day that this conversation took place.

i.d.t. – ninety

```
T W E N T Y
T W E L V E
  S E V E N
    N I N E
    N I N E
    N I N E
    N I N E
    F I V E
    F I V E
    F I V E
-----------
N I N E T Y
```

i.d.t. – eightynine

directed approach: page 90
solution: page 111

E	G	H	I	N	O	S	T	V	Y
0	1	2	3	4	5	6	7	8	9

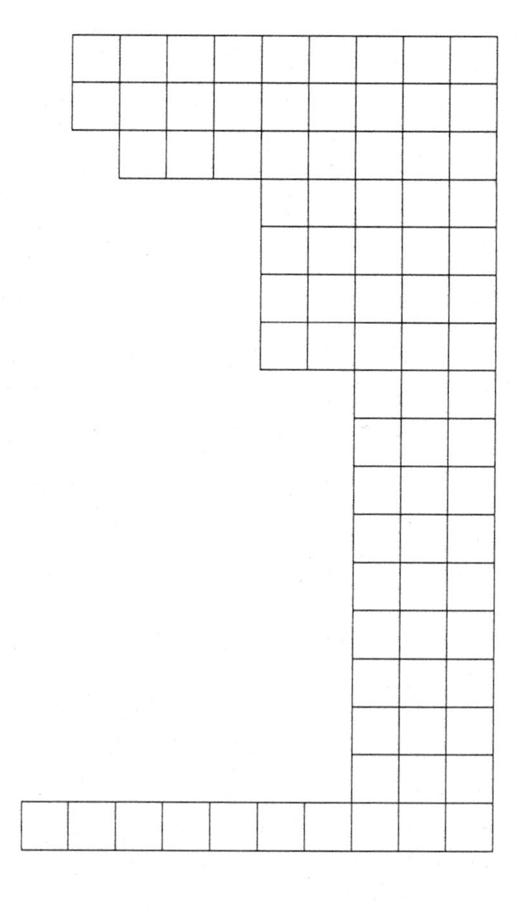

i.d.t. – eightynine

```
S E V E N T E E N
S E V E N T E E N
  E I G H T E E N
        S E V E N
        S E V E N
        S E V E N
        S E V E N
            O N E
            O N E
            O N E
            O N E
            O N E
            O N E
            O N E
            O N E
            O N E
---------------------
E I G H T Y N I N E
```

in a word
directed approach: page 91
solution: page 109

A	E	G	H	I	L	M	R	S	T
0	1	2	3	4	5	6	7	8	9

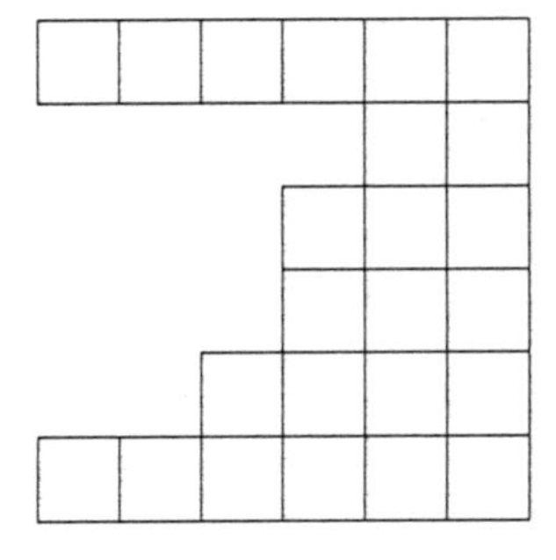

If an integer N can be expressed as a sum of squares, then so can 2N. For example, $25 = 3^2 + 4^2$ and $2(25) = 50 = 1^2 + 7^2$.

in a word

If you've ever been behind the wheel of a car and had your two children sitting behind you, then you've experienced the hostilities of warfare. Nothing brings out the venom in an adorable tyke as rapidly as the specter of sharing the back seat of an auto with his or her sibling.

There are three standard procedures currently available to defuse this type of explosive situation:

1. gags and/or restraints
2. tranquilizer darts
3. trigrams

Trigrams is a game invented by Lester, a close friend of mine, and it is by far the most humane of the listed alternatives. What's more, it's quite an enjoyable diversion, especially on a long and boring drive. Its playing pieces are license plates which display three letters and three numbers.

To illustrate, suppose you, the driver, spot a nearby vehicle sporting the license plate CHB 416. You identify the target to the rear seat combatants and the round commences. The first to come up with a legitimate English word that contains the letter combination CHB in it earns the prize of 416 points. Acceptable responses here would include watCHBand, switCHBlade, and hatCHBack, but CHuBby would not qualify, since the desired trigram cannot be interrupted by any intervening letters. The winner of the contest is the first to attain a pre-announced point total—5,000 is reasonable for a one-hour highway drive.

Through the years, Lester has amassed some fascinating trigrams, and he has generously allowed me to share sixteen of them with you. Some are more challenging than others, but rest assured—each and every one appears in at least one English word.

1. BEK
2. CCY
3. DHP
4. EUE
5. FGH
6. GNT
7. HIH
8. IKD
9. IOA
10. MSD
11. NTN
12. OKK
13. OUH
14. VEW
15. ZAA
16. ZOP

Try trigrams the next time the opportunity arises. You'll no doubt tip your hat to my friend and readily acknowledge that

L E S T E R
I S
T H E
T R I
G R A M
———————
M A S T E R .

fifty's not nifty

directed approach: page 92
solution: page 111

A	D	E	H	I	N	R	T	U	V
0	1	2	3	4	5	6	7	8	9

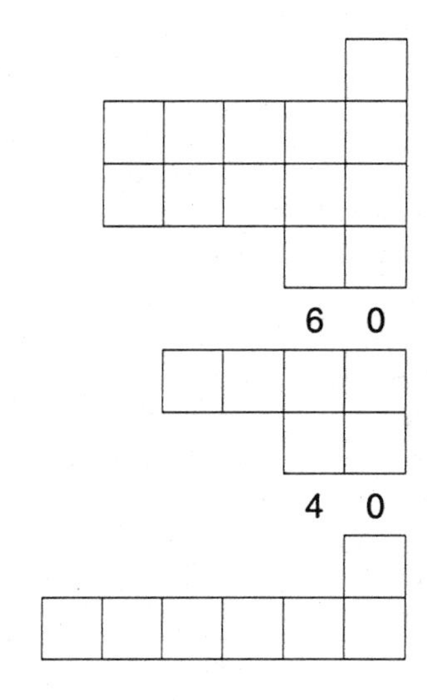

$$\frac{37^3 + 13^3}{37^3 + 24^3} = \frac{37 + 13}{37 + 24}$$

fifty's not nifty

Here's an old algebraic chestnut, recast in alphametic format. To be definite, assume that all average speeds are given in miles per hour.

Towns A and B are connected by a straight highway. I live in Town A but often travel to Town B on business, returning home later in the day. Since I can manage to avoid rush-hour traffic by leaving in the morning whenever I wish but cannot do so in the other direction, my average speed is much better traveling from A to B than from B to A. Find my average speed for the round-trip excursion if

```
          I
  D R I V E
  T H E R E
        A T
        6 0
    T H E N
        A T
        4 0
          I
-----------
R E T U R N .
```

Make the most of this DRIVE and feel free to reuse any digits that already appear.

Every integer of the form ABABAB is divisible by 7 and every integer of the form ABCABC is divisible by 11.

i.d.t. – hundred
directed approach: page 93
solution: page 117

D E H I N R S T U V

0 1 2 3 4 5 6 7 8 9

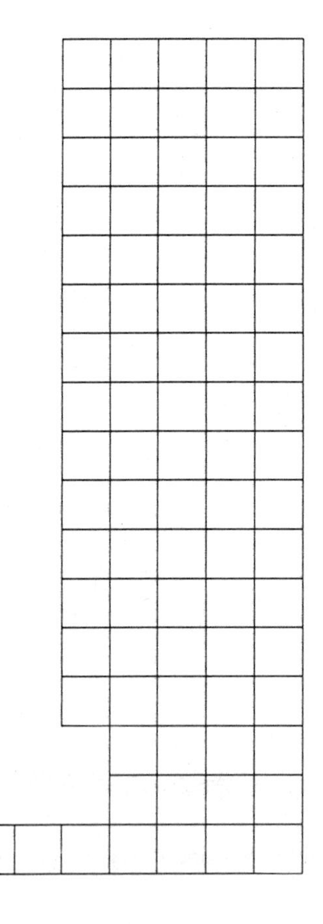

as a matter of fact
directed approach: page 94
solution: page 113

E G H N O R T U Y

0 1 2 3 4 5 6 7 8 9

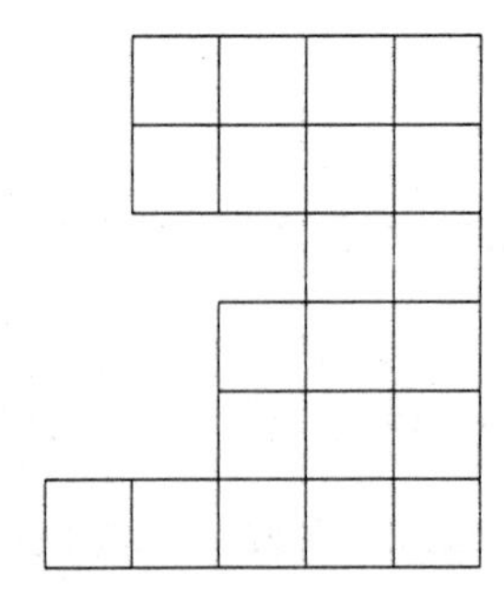

i.d.t. – hundred

```
    S E V E N
    S E V E N
    S E V E N
    S E V E N
    S E V E N
    S E V E N
    S E V E N
    S E V E N
    S E V E N
    S E V E N
    T H R E E
    T H R E E
    T H R E E
    T H R E E
      N I N E
      N I N E
  -----------
H U N D R E D
```

as a matter of fact

Optical illusions provide simple and direct evidence to support the old adage that appearances can be deceiving. This maxim is a valid one in the world of mathematics as well.

Using the nomenclature of the Romans, study the seemingly false claim that

$$XI + I = X$$

Without introducing, deleting, or shifting any symbols whatsoever, this statement can be legitimatized! Others have tried and failed—now it's

```
 Y O U R
 T U R N
     T O
   G E T
   T H E
---------
T R U T H ,
```

where an odd TURN is required for the desired solution.

rhyme time

directed approach: page 94
solution: page 114

C	E	I	L	M	N	O	P	S	T
0	1	2	3	4	5	6	7	8	9

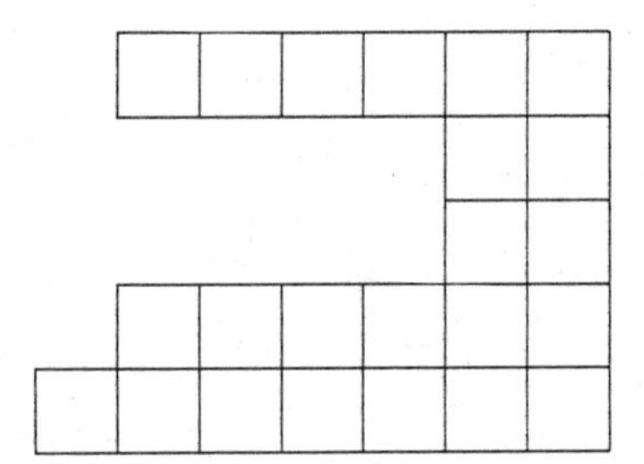

rhyme time

Marriage, driving, and hunting are three seemingly unrelated activities that share a common prerequisite—the need for a special permit to participate. In a similar fashion, those who pen sonnets, odes, haikus, limericks, and the like require guidelines to ply their trade as well. What form does this unofficial permission take?

S I M P L E –
I T
I S
P O E T I C
L I C E N S E !

Writing poetry may appear to be a rather daunting task, but there is a poem lurking within each of us. To verify this assertion, let us explore a very rudimentary structure that is appropriately called "terse verse." As this label illustrates, members of this category consist of exactly two words that just happen to rhyme. Put another way, each is the shortest possible example of a "witty ditty." Clues have been provided below for a dozen more terse verses. (Note that the first clue is itself a terse verse!) After constructing these, try composing some others on your own. Who knows—some day, you could be the author of your very own "poem tome."

1. fat cat
2. car mishap
3. counterfeit coin
4. astute lawyer
5. hometown boy
6. monster movie
7. quadrillion
8. sales talk
9. ebb tide
10. emotional butler
11. Acapulcan dictionary
12. Venezuelan rattles

i.d.t. – ninetyeight
directed approach: page 95
solution: page 110

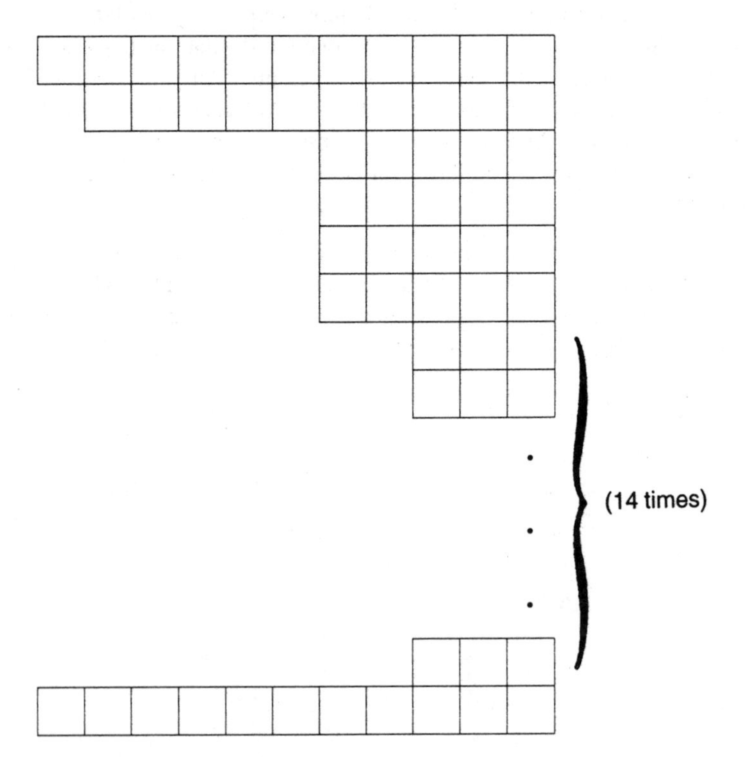

i.d.t. – ninetyeight

```
T W E N T Y T H R E E
  T W E N T Y N I N E
            E I G H T
            E I G H T
            E I G H T
            E I G H T
                O N E ⎫
                O N E ⎪
                    . ⎪
                    . ⎬ (14 times)
                    . ⎪
                O N E ⎭
---------------------
N I N E T Y E I G H T
```

The following 70-digit number is prime:

1234567891234567 89 . . . 1234567891234567

---

identity crisis

directed approach: page 96
solution: page 109

A B C E L M S

1 3 4 6 7 8 9

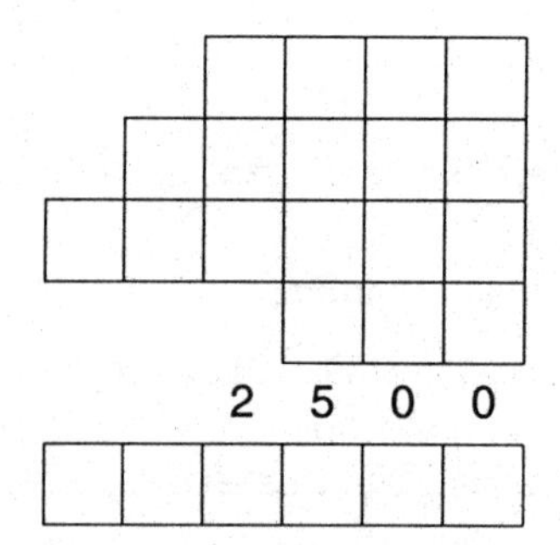

identity crisis

The AC/DC Electrical Supply Company keeps three types of three-wire cables in stock. The first type consists of three black wires, the second consists of three white wires, and the third consists of two black wires and one white wire.

A recent shipment of 2,500 cables was delivered to AC/DC in three separate cartons, each containing one type of cable exclusively. Unfortunately, the cable manufacturer mixed up the descriptive labels and placed an incorrect label on each of the three cartons. When Mr. A. M. Pere, president of AC/DC, was informed about the improper identification, his plan was to open any two of the cartons, note the contents of each, and thus learn what is truly in each of the cartons.

"Wait!" exclaimed his executive secretary, Mabel. "I need to open just one of the cartons to determine the actual contents of all three."

After a moment's thought, Mr. Pere realized that his secretary's claim was entirely justified. Can you explain how

$$\begin{array}{r} \text{A B L E} \\ \text{M A B E L} \\ \text{L A B E L S} \\ \text{A L L} \\ 2\ 5\ 0\ 0 \\ \hline \text{C A B L E S} \end{array},$$

without reusing the digits 0, 2, and 5?

For prime p, define p primorial, denoted by $p^{\#}$, to be the product of all primes $\leq$ p. (Thus, $5^{\#} = 2 \times 3 \times 5 = 30$.) The largest known primorial that can be expressed as a product of two consecutive integers is $17^{\#}$, which equals $2 \times 3 \times 5 \times 7 \times 11 \times 13 \times 17$ or 510510. This number is the product of 714 and 715.

$$123789 + 561945 + 642864 = 242868 + 323787 + 761943$$

and

$$123789^2 + 561945^2 + 642864^2 = 242868^2 + 323787^2 + 761943^2$$

If a, b, c, d, e, f, g, and h represent any eight consecutive integers, then

$$a^2 + d^2 + f^2 + g^2 = b^2 + c^2 + e^2 + h^2$$

generation gap

Three acts that undoubtedly evoke grimaces from the general populace are sucking on a lemon, scratching a piece of chalk on a blackboard, and trying to solve a verbal problem using algebraic techniques. The most violent reaction very often comes from the last of these, owing to the fact that most people are, to a certain degree, mathphobic. When viewed with objectivity, however, algebra is revealed to be a logical power tool that can unravel seemingly complex and knotty situations.

The embryonic stage of this tool dates back to the early part of the ninth century when Al-Khowarizmi, an Arabic mathematician, wrote *Al-jabr wa'l Muqabalah.* (Loosely translated, the title means "restoration and balancing.") The fundamental principle contained therein asserted that whenever an operation is performed upon one member of an equation, that very same operation must be performed upon its other member in order to "restore" equality and thus maintain "balance." It would not be until the late sixteenth century that the use of letters to represent unknown quantities would actually be introduced, but the seed was planted. Through the years, this little acorn has most certainly grown into a majestic oak!

Two rather complicated-sounding problems serve to illustrate the potency of algebra.

1. A rope over the top of a fence has the same length on each side and weighs 1/3 pound per foot. On one end hangs a monkey holding a banana and on the other end hangs a weight equal to the weight of the monkey. The banana weighs 2 ounces per inch. The rope is as long as the age of the monkey, and the weight of the monkey (in ounces) is as much as the age of the monkey's mother. The sum of the ages of the monkey and the monkey's mother is 30 years. When one-half the weight of the monkey is added to the weight of the banana, the total is one-fourth as much as the sum of the weights of the weight and the rope. The monkey's mother is one-half as old as the monkey will be when it is three times as old as its mother was when she was one-half as old as the monkey will be when it is as old as its mother will be when she is four times as old as the monkey was when it was twice as old as its mother was when she was one-third as old as the monkey was when it was as old as its mother was when she was three times as old as the monkey was when it was one-fourth as old as it is now. How long is the banana?

generation gap

directed approach: page 96
solution: page 112

F H I N S T W Y

0 1 2 3 4 5 6 7 8 9

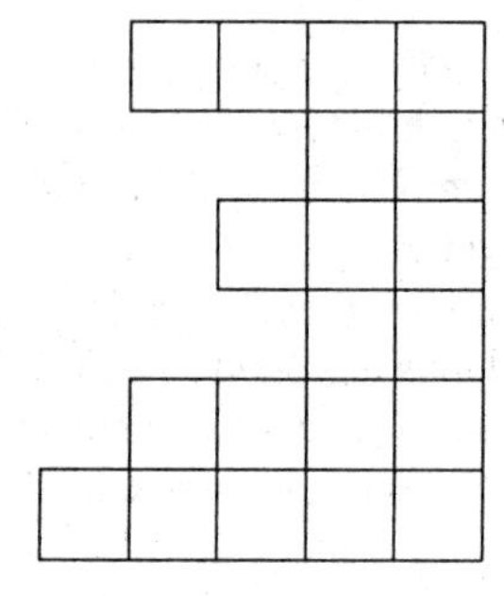

The nth triangular number, denoted by T_n, is given by $T_n = \frac{1}{2}n(n + 1)$. The first ten such numbers are 1, 3, 6, 10, 15, 21, 28, 36, 45, and 55. Triangular numbers 15 and 21 are the smallest ones with the property that their sum and difference are also triangular numbers.

Solution:

Let x = the weight of the weight (in ounces)
= the weight of the monkey (in ounces)
= the age of the monkey's mother (in years)
y = the age of the monkey (in years)
= the length of the rope (in feet)
z = the weight of the banana (in ounces).

Then,

$$x + y = 30,\ \tfrac{1}{2}x + z = \tfrac{1}{4}\left(x + \tfrac{16}{3}y\right),\ \text{and}$$

$$x = \tfrac{1}{2}\left(3\left(\tfrac{1}{2}\left(4\left(2\left(\tfrac{1}{3}\left(3\left(\tfrac{1}{4}y\right)\right)\right)\right)\right)\right)\right).$$

After simplification, we obtain

$$x + y = 30,\ 3x + 12z = 16y,\ \text{and}\ x = \tfrac{3}{2}y.$$

Solving the first and third equations simultaneously, we find $x = 18$ and $y = 12$. Substituting these values into the second equation gives $z = 11.5$. Since the banana weights 2 ounces per inch, the length of the banana is $11.5 \div 2 = 5.75$ inches. (WHEW!)

2. Jack and his father and John and his father went fishing. At the end of the day, the fathers had caught twice as many fish as had the sons. Further, Jack's father had caught twice as many fish as had John's father. If 35 fish were caught in all, who is older, Jack or John?

If the first problem apparently had too much to say, this one seems to have too little. Moreover, there is absolutely no reference to the ages of the individuals until the question is posed. Yet the problem is solvable! Analyze the described situation with care and then call upon Al Khowarizmi's legacy.

$$\begin{array}{r} \text{T H I S} \\ \text{I S} \\ \text{W H Y} \\ \text{I T} \\ \underline{\text{I S N' T}} \\ \text{F I S H Y} \end{array}.$$

In fact, the least FISHY should be your goal here.

10^{33} is expressible as the product of two factors, neither of which has a zero digit:

$$8589934592 \times 116415321826934814453125$$

$$\frac{1}{3} = \frac{1+3}{5+7} = \frac{1+3+5}{7+9+11} = \frac{1+3+5+7}{9+11+13+15} = \frac{1+3+5+7+9}{11+13+15+17+19} = \cdots$$

$$95000^4 + 217519^4 + 414560^4 = 422481^4$$

all mixed up

According to Webster, an anagram is a transposition of the letters of a word or phrase to form another word or phrase related in meaning to the original. Logophiles have noted that the letters of the word ANAGRAMS can be rearranged into ARS MAGNA, the Latin phrase for "great art." As evidence of this greatness, consider the two dozen examples that follow. Some were culled from the vast listing found in "Palindromes and Anagrams," Howard Bergerson's book published in 1973 by Dover Publications.

1. EVIL / VILE
2. ANGERED / ENRAGED
3. ASTRONOMER / MOONSTARER
4. MEASURED / MADE SURE
5. PITTANCE / A CENT TIP
6. ENDEARMENT / TENDER NAME
7. DECLARATION / AN ORAL EDICT
8. DISINTEGRATION / DARN IT, IT IS GONE
9. PRESBYTERIANS / BEST IN PRAYERS
10. THE EYES / THEY SEE
11. INCOME TAXES / EXACT MONIES
12. GOLD AND SILVER / GRAND OLD EVILS
13. GRAND FINALE / A FLARING END
14. ELEVEN + TWO / TWELVE + ONE
15. THE NUDIST COLONY / NO UNTIDY CLOTHES
16. ADOLF HITLER / HATED FOR ILL
17. THE ACTIVE VOLCANOS / CONES EVICT HOT LAVA
18. THE STATUE OF LIBERTY / SOFT-LIT BEAUTY THERE
19. CIRCUMSTANTIAL EVIDENCE / ACTUAL CRIME ISN'T EVINCED
20. MIGUEL CERVANTES DE SAAVEDRA / GAVE US A DAMNED CLEVER SATIRE
21. THE AMERICAN INDIAN RESERVATION / IT IS ONE AREA RED MAN CAN THRIVE IN
22. IVANHOE BY SIR WALTER SCOTT / A NOVEL BY A SCOTTISH WRITER
23. WASHINGTON CROSSING THE DELAWARE / HE SAW HIS RAGGED CONTINENTALS ROW
24. THE U.S. LIBRARY OF CONGRESS / IT'S ONLY FOR RESEARCH BUGS

all mixed up
directed approach: page 97
solution: page 118

A B C E G L M N R S

0 1 2 3 4 5 6 7 8 9

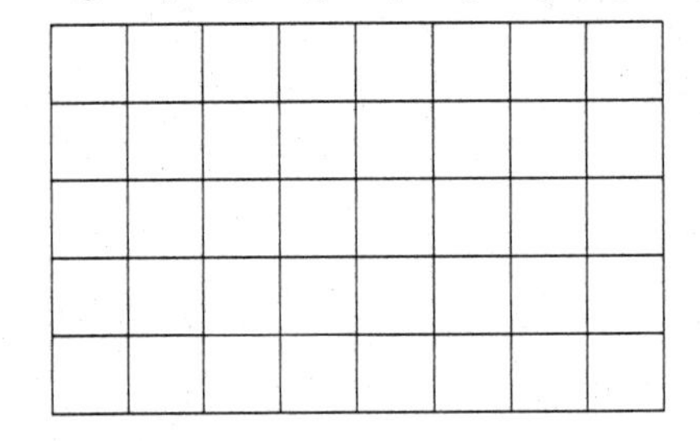

i.d.t. – ninetynine
directed approach: page 98
solution: page 113

E F H I N R S T V Y

0 1 2 3 4 5 6 7 8 9

The trick in constructing a good illustration of this genre is revealed alphametically:

```
A N A G R A M S
A N A G R A M S
A N A G R A M S
A N A G R A M S
---------------
S C R A M B L E .
```

Lest we forget, here's your homework assignment—use your newly-honed scrambling skills to make one word out of the letters of RODE NOW.

i.d.t. – ninetynine

```
    S E V E N T E E N
    S E V E N T E E N
    S E V E N T E E N
          F I F T E E N
              T H R E E
              T H R E E
              T H R E E
              T H R E E
              T H R E E
                N I N E
                N I N E
-------------------------
N I N E T Y N I N E
```

fourplay

directed approach: page 99
solution: page 111

C E F L O R S U Y

1 2 3 4 5 6 7 8 9

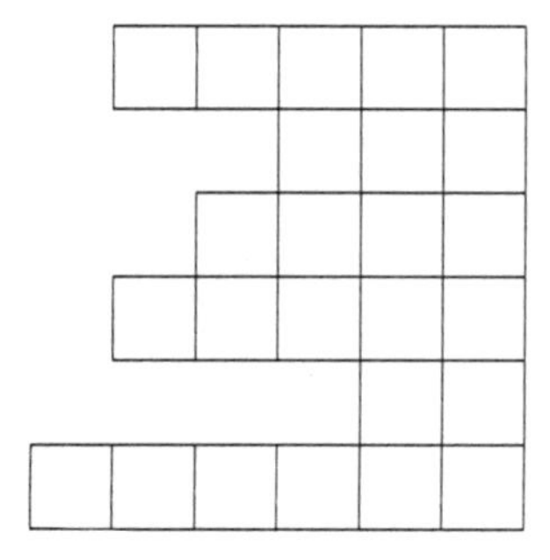

The sum of the cubes of 11, 12, 13, and 14 is equal to the cube of 20.

$$1! + 6! + 9! = 36301$$
$$3! + 6! + 3! + 0! + 1! = 1454$$
$$1! + 4! + 5! + 4! = 169$$

---

fourplay

Everyone knows that four fours make sixteen. However, if we broaden our horizons by expanding our list of operations and symbols, then the same four fours can make much, much more. For instance, we can write

$$71 = (4! + 4.4) \div .4$$
$$95 = 4(4!) - 4 \div 4$$
$$134 = 4! + 44 \div .4$$

and

$$168 = (4! + 4)(\sqrt{4} + 4).$$

Let us agree to supplement the ordinary arithmetic operations with concatenation and exponentiation and also permit the use of decimal points, factorials, radical signs, and parentheses. Adhering to these guidelines, why not try your hand at the production process? In particular, see if you can manufacture representations for 89, 104, 109, 127, 142, and 150. In each case,

```
 Y O U' L L
       U S E
     F O U R
   F O U R S ,
         O F
 ___________
 C O U R S E
```

and you'll refrain from using the digit "0" in the solution of the alphametic.

$$312 \times 221 = 68952\ ;\ \ 213 \times 122 = 25986$$
$$123 \times 102 = 12546\ ;\ \ 321 \times 201 = 64521$$

---

i.d.t. – sixtysix
directed approach: page 100
solution: page 117

E	H	I	N	O	R	S	T	X	Y
0	1	2	3	4	5	6	7	8	9

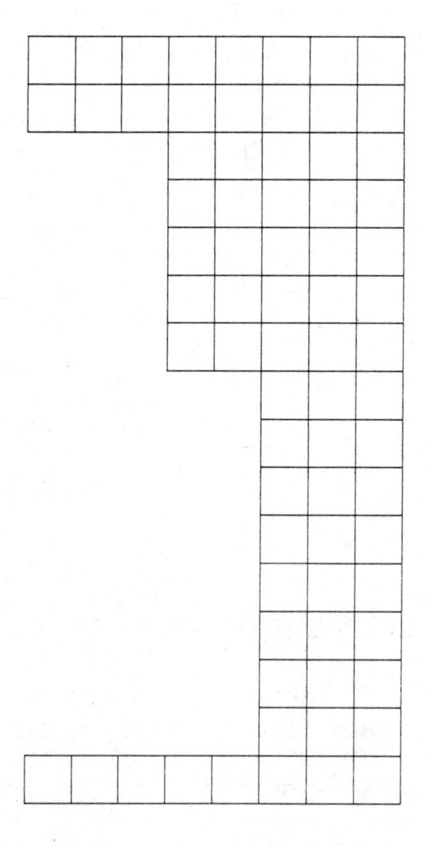

i.d.t. – sixtysix

```
N I N E T E E N
N I N E T E E N
      T H R E E
      T H R E E
      T H R E E
      T H R E E
      T H R E E
          S I X
          O N E
          O N E
          O N E
          O N E
          O N E
          O N E
          O N E
---------------
S I X T Y S I X
```

$$73 \times 9 \times 42 = 7 \times 3942$$

i.d.t. – ninetytwo

directed approach: page 101
solution: page 109

E	F	H	I	N	O	R	T	W	Y
0	1	2	3	4	5	6	7	8	9

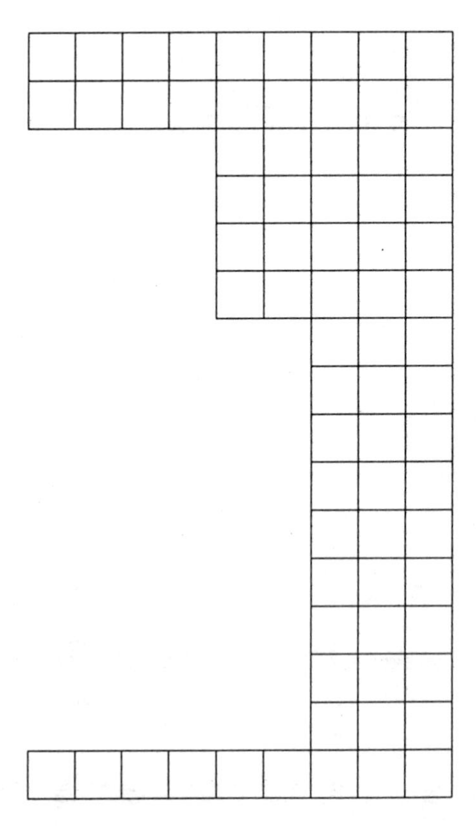

i.d.t. – ninetytwo

```
F O R T Y N I N E
T W E N T Y O N E
        T H R E E
        T H R E E
        T H R E E
        T H R E E
            T W O
            O N E
            O N E
            O N E
            O N E
            O N E
            O N E
            O N E
            O N E
-----------------
N I N E T Y T W O
```

$$(32)^3 + 2(34)^3 + 2(65)^3 + (67)^3 = (76)^3 + 2(56)^3 + 2(43)^3 + (23)^3$$

whatchamacallits

directed approach: page 102
solution: page 113

A	C	D	E	H	I	L	S	T	W
0	1	2	3	4	5	6	7	8	9

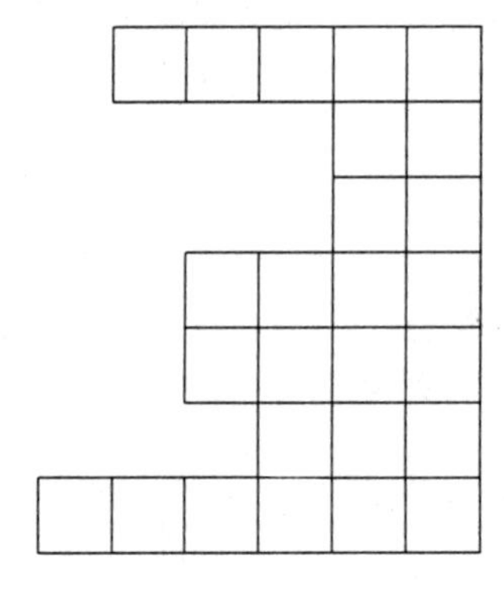

$$\frac{143185}{1701856} = \frac{1435}{17056}$$

whatchamacallits

Grammarians subdivide the set of nouns into two categories—those that are proper and those that are common. Some members in the latter grouping, however, are anything but common, in spite of the fact that they are synonymous with well-known things. An excursion through an unabridged dictionary yields fascinating illustrations of what might more appropriately be labeled as "uncommon nouns." A sampling of these is displayed alphabetically in Column A. See how many you can match with their more mundane definitions which appear somewhere in Column B.

At some point in the proceedings, you'll undoubtedly be heard to exclaim

THAT'S
IT ?
IS
THAT
WHAT
IT'S
CALLED ??

COLUMN A	COLUMN B
1. aglet	a. frivolous or giddy girl or woman
2. barm	b. small cavity in a rock
3. biggin	c. person who explores caves
4. bight	d. froth or head on a glass of beer
5. bissextus	e. palm of the hand
6. chad	f. hatchetlike tool used for cutting holes in slate
7. fizgig	g. container in a coffeepot that holds the grounds
8. hallux	h. the day added to the calendar each leap year
9. hirci	i. slash used in grammar to separate alternatives
10. loof	j. sheath on the end of a shoelace to simplify a pass through the holes
11. lunule	k. small circular piece of paper that results from using a hole-puncher
12. spelunker	l. armpit hair
13. virgule	m. middle or slack part of a rope
14. vug	n. white area at the base of a fingernail
15. zax	o. big toe

home sweet home

directed approach: page 102
solution: page 117

C	E	H	I	N	O	S	T	U	W
0	1	2	3	4	5	6	7	8	9

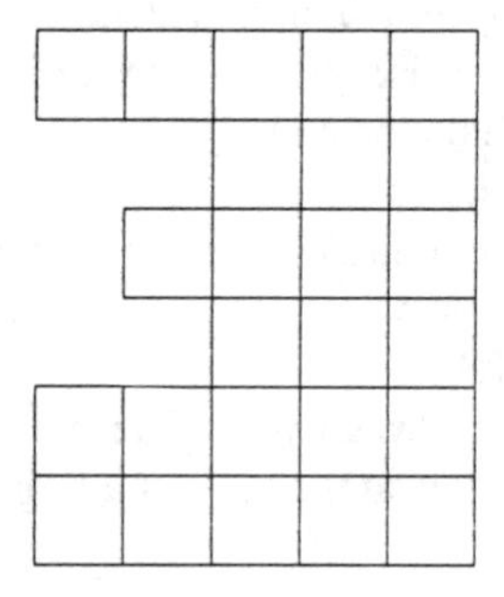

home sweet home

An American, a Canadian, an Englishman, a Frenchman, and a Russian each own one of five houses on a block.

1. The Englishman owns the red house.
2. The Russian has a dog.
3. The man who owns the green house drinks rum.
4. The Frenchman drinks scotch.
5. The green house is immediately to the right of the white house.
6. The man who drives a Cadillac has a cat.
7. The man who drives a Buick owns the yellow house.
8. The man who owns the middle house drinks bourbon.
9. The American owns the first house.
10. The man who drives a Ford lives next door to the man who has a parrot.
11. The man who drives a Buick lives next door to the man who has a horse.
12. The man who drives a Chevrolet drinks gin.
13. The Canadian drives a Pontiac.
14. The American lives next door to the blue house.

If each man has one pet, drives one make of car, and drinks one type of liquor, which one owns the turtle, which one drinks vodka, and

W H I C H
O N E
O W N S
T H E
W H I T E
H O U S E ?

i.d.t. – ninetysix
directed approach: page 103
solution: page 118

E	G	H	I	N	R	S	T	X	Y
0	1	2	3	4	5	6	7	8	9

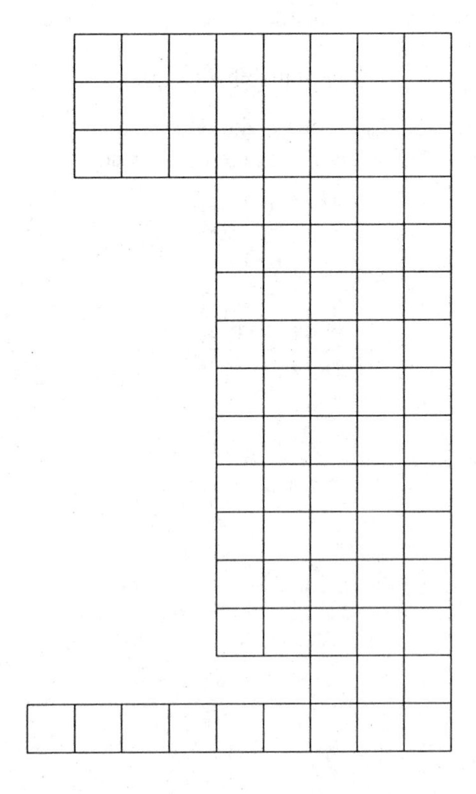

i.d.t. – ninetysix

```
 N I N E T E E N
 N I N E T E E N
 E I G H T E E N
       T H R E E
       T H R E E
       T H R E E
       T H R E E
       T H R E E
       T H R E E
       T H R E E
       T H R E E
       T H R E E
       T H R E E
           T E N
-----------------
N I N E T Y S I X
```

$$(1111111111)^2 = 1234567900987654321$$

ready, set, go
directed approach: page 103
solution: page 114

A	E	G	H	I	L	M	N	S	T
0	1	2	3	4	5	6	7	8	9

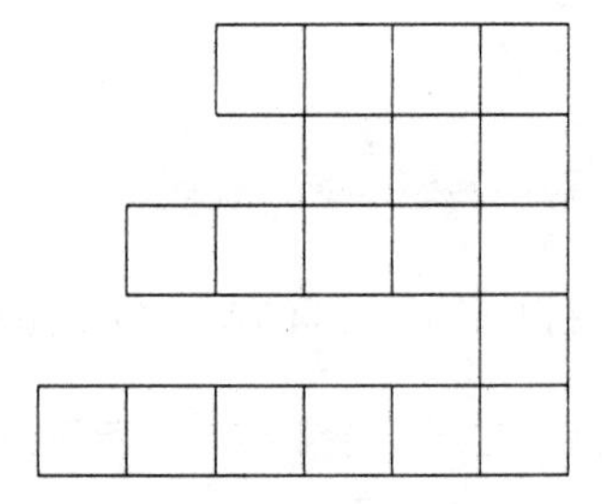

ready, set, go

We've all experienced the frustration that results when commonly-known facts "slip our mind." Each December, for instance, we're reminded that Santa Claus is coming to town, led by his eight reindeer—Dasher, Dancer, Prancer, Vixen, Comet, Cupid, Donder, and Blitzen. Listing the members of this familiar octet, however, might prove to be somewhat elusive were it not for the introductory bars of "Rudolph, The Red-Nosed Reindeer."

The little quiz that follows will test your knowledge of a dozen sets of objects. How many of these rosters can you write down without hesitation?

1. Name all THREE musketeers.
2. Name all FOUR voices in a choir.
3. Name all FIVE Great Lakes.
4. Name all SIX presidents featured on U.S. coins.
5. Name all SEVEN dwarfs.
6.

```
   N A M E
     A L L
 E I G H T
       'N'
-----------
S T A T E S .
```

(Appropriately, N should be equal to 8.)

7. Name all NINE planets in the solar system.
8. Name all TEN Canadian provinces.
9. Name all ELEVEN states in the Confederacy.
10. Name all TWELVE signs of the zodiac.
11. Name all THIRTEEN original colonies.
12. Name all FOURTEEN National League baseball teams.

i.d.t. – eightytwo
directed approach: page 104
solution: page 110

E	G	H	I	N	O	R	T	W	Y
0	1	2	3	4	5	6	7	8	9

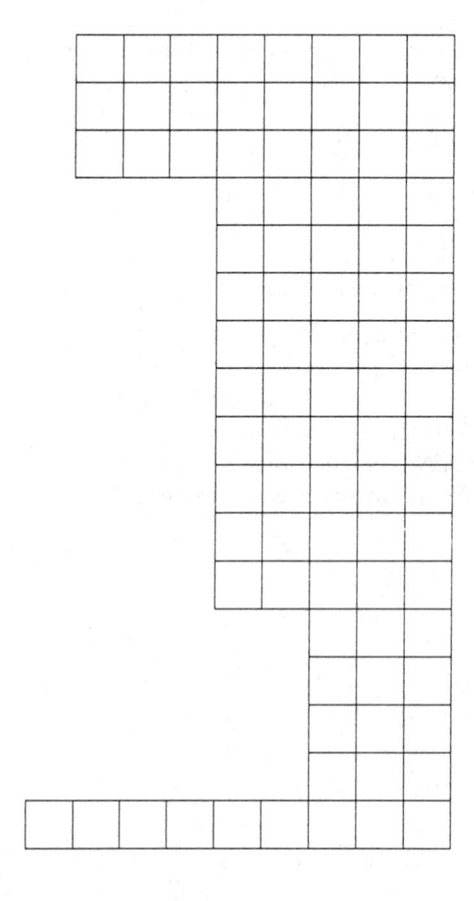

i.d.t. – eightytwo

```
T H I R T E E N
T H I R T E E N
T H I R T E E N
        T H R E E
        T H R E E
        T H R E E
        T H R E E
        T H R E E
        T H R E E
        T H R E E
        T H R E E
        T H R E E
              T E N
              T W O
              T W O
              T W O
_________________
E I G H T Y T W O
```

nothing but the truth
directed approach: page 105
solution: page 116

A	C	G	I	L	M	N	O	S	T
0	1	2	3	4	5	6	7	8	9

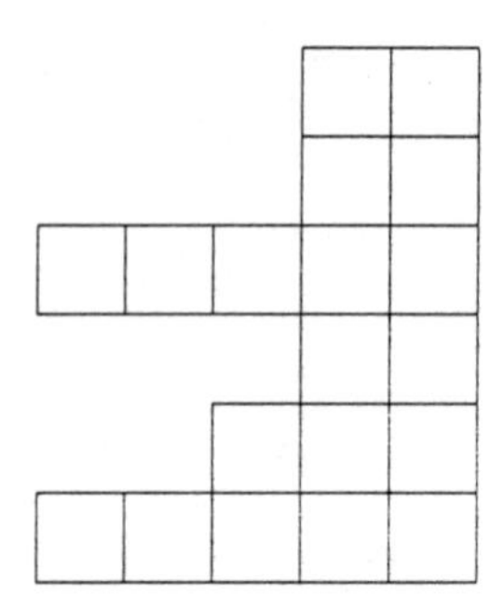

nothing but the truth

One and only one of the following statements is true:

1. Exactly one of these statements is false.
2. Exactly two of these statements are false.
3. Exactly three of these statements are false.
4. Exactly four of these statements are false.
5. Exactly five of these statements are false.
6. Exactly six of these statements are false.
7. Exactly seven of these statements are false.
8. Exactly eight of these statements are false.

To uncover the solitary true statement, what tool do you use?

```
      I S
      I T
M A G I C ?
      N O ,
    I T' S
-----------
L O G I C !
```

For efficiency, the most LOGIC is desirable.

? ! ? ! ? ! ? ! ? ! ? ! ? ! ? ! ? ! ? ! ?

SECTION 2

DIRECTED APPROACHES

? ! ? ! ? ! ? ! ? ! ? ! ? ! ? ! ? ! ? ! ?

cover

```
T A K E
      A
L O O K
    A T
      A
G O O D
-------
B O O K
```

A joint examination of the tens and hundreds columns produces the value of A + O. From here, the tens column permits only one assignment for the letter K when the carryover from the units column is taken into consideration. There are but two available values for B, which appears in the thousands column of the sum. Keeping in mind that no leading digit can equal zero, each choice of B yields a unique trio of digits for T, L, and G. In both instances, the previously-established value of A + O now gives a specific pair of values for A and O. Shifting to the units column, an equation involving A, T, E, and D can be found. From the earlier part of the analysis, exactly one of the options for A and T determines an interchangeable pair of digits for the letters E and D. Once T has been selected, the two remaining members of the aforementioned trio of digits correspond with L and G in an interchangeable manner as well.

dedication

```
  E U L E R
  G A U S S
-----------
G R E A T S
```

Determination of the values of R and G is immediate from the units column and hundred thousands column, respectively. This leaves only one viable option for the value of E when the ten thousands column is examined. Once E has been assigned, the carryover into the hundreds column follows next. The constraint suggests that A should be chosen from the pool of remaining digits to be as large as possible, while still allowing satisfaction of the equation that results from the thousands column. With A selected in this manner, the values of U and L emerge from consideration of the thousands and hundreds columns. Finally, the tens column yields the values of S and T.

preface

```
Y O U R S
    V E R Y
-----------
T R U L Y
```

The value of S is obvious from the units column, with the assignment for E following immediately afterward when the hundreds column is examined. Joint consideration of the tens and thousands columns now admits only one value of R, which in turn produces the value of L (from the tens column) and an interchangeable pair of digits for O and V (from the thousands column). Based upon the constraint, just one option is available for the value of Y, with the determination of the value of T coming next from the ten thousands column. The one remaining digit is finally assigned to U.

greetings

```
      S H E
  S H A L L
  S H A K E
          4
-----------
  H A N D S
```

The ten thousands column and the carryover into it produce three feasible assignments for S and H. Noting the unavailability of the digit "4," each of these gives a value for A when the hundreds and thousands columns are jointly considered. In only two cases, however, will the hundreds column generate an acceptable value for N. Examine the units column in both instances to establish a relationship between E and L and satisfy it with remaining digits wherever possible. Exactly one of these options offers an appropriate choice for the values of K and D upon inspection of the tens column.

i.d.t. – fiftytwo

```
  T H I R T E E N
    F I F T E E N
    F I F T E E N
            T W O
            T W O
            T W O
            T W O
            O N E
  ---------------
  F I F T Y T W O
```

List possible assignments for T and F by considering the leftmost column, thus establishing the carryover into the hundred thousands column. From that column and then the millions column, obtain I and H, respectively, wherever the needed digits are accessible. Find R from the ten thousands column and values that can correspond to Y from the thousands column. Move to the units column to interrelate N, O, and E and choose digits that satisfy the resulting condition as availability permits. The remaining digit in one case proves to be the correct value for W when the tens column and the ensuing carryover are examined.

i.d.t. – eightysix

```
  S E V E N T E E N
        S I X T E E N
        S I X T E E N
        S I X T E E N
            S E V E N
            S E V E N
            S E V E N
  -----------------
  E I G H T Y S I X
```

The two columns at the far left admit a pair of possibilities for the values of E and S. One of these is quickly eliminated by finding the minimal value of I from the tens column and noting that it cannot be attained in the ten millions column. In the remaining case, the same two sources yield the choice for I. An explicit value for the carryover into the tens column is now available, and this generates the values of N and X upon examination of the units column. Move next to the hundreds column to obtain the digit that corresponds to V, and use the conclusion to select G from the millions column. Only one option exists for the value of H when the hundred thousands column is scrutinized. Lastly, establish the link between T and Y from the thousands column. Note that the two unassigned digits satisfy it and at the same time produce the correct carryover into the ten thousands column.

threesomes

```
  S T R A N G E
  S T R A N G E
  S T R A N G E
---------------
T R I P L E T S
```

The millions column in conjunction with the initial condition allows only three possible selections for the value of S, one of which is immediately eliminated when the units column is checked. The two remaining values both give the same value of T upon examination of the leftmost column. Go to the units column again, this time to determine the value of E, and move leftward to obtain values of G and N from the tens and hundreds columns, respectively. At this juncture, there is a single case to pursue. Reconsideration of the millions column yields the value of R, which leads to the value of I from the hundred thousands column. The relationship involving A and L obtained from the thousands column can now be satisfied in just one way with the digits that remain. This choice generates the appropriate carryover into the ten thousands column to permit the matching of the sole unassigned digit with P.

i.d.t. – fourhundredfortyone

```
F O U R H U N D R E D F O U R T E E N
                            T H R E E
                              F O U R
                              F O U R
                              F O U R
                                T E N
                                O N E
                                O N E
-------------------------------------
F O U R H U N D R E D F O R T Y O N E
```

The parity of R can be found from the units column at the outset, leading to just two choices for its actual value when the ten thousands column is examined. In each instance, the hundred thousands column yields the corresponding selection for U. When the units and tens columns are viewed in tandem, one of the avenues gives two possible values for the pair E and N, while the other proves to be a dead end. Go next to the hundreds column to designate options for T and O. In only one case does consideration of the thousands column give rise to an equation involving H, F, and Y that can be made true with still-available digits. Conclude by matching the lone unassigned digit with D.

from a to z

```
  H O W
M U C H
W I L L
  O U R
W O R M
-------
M U N C H
```

Beginning at the left, acknowledge the constraint by selecting the smaller of the two possible values for M. The same rationale dictates the minimal choice for the value of U when the thousands column is investigated. This column then offers two options for the value of W. In each case, the units column produces the value of the sum L + R, which in turn gives the assignment for O upon consideration of the tens column. At this point, the required carryover into the thousands column proves to be unattainable in one of the cases, leaving a single avenue to pursue. Use the hundreds column to establish the equation involving I, H, and N and find the value of N from it. Moreover, the same source yields one pair of digits as associates for I and H. A review of the sum L + R now allows but one interchangeable pair of available digits corresponding to these letters, leaving only one digit for the value of C. Still another glance at the initial condition fixes the value of H and thus the value of I.

i.d.t. – sixtynine

```
S E V E N T E E N
  N I N E T E E N
        T H R E E
            T E N
            T E N
            T E N
-----------------
S I X T Y N I N E
```

Observing the upper bound imposed upon E + N from the ten millions column, use the contents of the units and tens columns to find the precise values of E and N. Joint consideration of the ten thousands and hundred thousands columns gives T next, thus leaving only one choice for the value of I when the ten millions column is re-examined. The equation that results from the millions column has a unique solution, thereby securing the assignments for both V and X. The values of R, H, and Y then follow from the hundreds, thousands, and ten thousands columns, respectively, leaving one digit available as the correspondent of S.

i.d.t. – eighty

$$12(\text{FIVE}) + 20(\text{ONE}) = \text{EIGHTY}$$

Three ten-shifts convert the problem into

```
  F I V E
    F I V E
    F I V E
    O N E
    O N E
-----------
E I G H T Y ,
```

from which the leftmost column clearly implies the value of E. The units column then gives Y directly, leaving only one viable choice for I when the ten thousands column is analyzed. Obtain the relationship involving V and T from the tens column and list the digits that make it so. In each case, move to the hundreds column and repeat the process, this time linking the letters N and H. Whenever satisfaction of this condition occurs, examine the thousands column to determine what must be true concerning the letters F, O, and G. When appropriate digits are available to meet this requirement, consider the carryover into the ten thousands column to distinguish the specific value of F, finding this possible in only one manner.

pun, pun, pun

```
  S O N S
R A I S E
  M E A T
---------
T H E R E
```

The constraint implies that the largest value of S should be examined first, with the value of T subsequently following from the units column. Consideration of the ten thousands column then produces assignments for R where available. The tens and hundreds columns next yield values for the sums N + A and O + I, respectively. Fulfill these conditions if possible, selecting O to be greater than I in deference to the requirement of maximality. Use the thousands column to find correspondents for M and H once A is fixed. This in turn establishes the value of N and also gives E by the process of elimination.

i.d.t. – seventy

```
  T W E N T Y
  E L E V E N
  E L E V E N
    T H R E E
    T H R E E
    T H R E E
    T H R E E
    T H R E E
    T H R E E
        T E N
-------------
S E V E N T Y
```

The units and tens columns together offer possibilities for the values of E and N and also establish the parity of T when the hundreds column is considered. In each instance, the hundred thousands column gives T explicitly and also leads to the assignment for S from the leftmost column. Relate V and R from the hundreds column and satisfy this relationship with available digits after noting the required parity of the carryover into the thousands column. This same column yields the value of H in two of the existing cases, but only one generates a solvable equation involving W and L from the ten thousands column. Associate the remaining digit with Y.

on a count

```
  T O T A L
  T H E S E
    D E L S
      A N D
-----------
D E L T A S
```

With the value of D apparent from the leftmost column, move one column to the right to list options for T and E. For each, the units column then gives the assignment for L, which subsequently generates the value of the sum S + N from the tens column. Continue leftward to establish the value of A from the hundreds column where available. Backtrack to designate possible choices for S and N while paying heed to the initial condition. In only one case will an interchangeable pair of digits still be present that is compatible with the condition imposed upon O and H by the thousands column.

array, array

```
    F I X
  V I C'S
    S I X
---------
P I C K S
```

The assignments for P, V, and I are immediately forthcoming from the thousands and ten thousands columns. Examination of the units column then produces the value of X, and the resulting carryover into the tens column establishes a relationship between C and K that can be satisfied in four different ways by the digits that are still available. Two of these can be eliminated when the hundreds column is considered. Of the two that remain, just one offers values of F and S where the latter is odd, as required.

i.d.t. – ninetyone

```
T W E N T Y T W O
  F O U R T E E N
  F O U R T E E N
    F I F T E E N
    F I F T E E N
              T W O
              T W O
              T W O
              T W O
              T W O
              O N E
-----------------
N I N E T Y O N E
```

The thousands column reveals that two choices exist for T. Joint consideration of the leftmost and units columns gives specific values for N and O when the latter column is analyzed. Only one of the two cases gives a carryover into the tens column that has the proper parity. Looking at this column and the carryover into the hundreds column is sufficient to produce assignments for both E and W. Taking into account the parities of the carryovers into the hundred thousands and millions columns, find explicit values for the sums F + R and I + U, respectively. Then the millions column yields the value of F, which subsequently generates the value of I when the ten millions column is checked. These selections give the values of R and U in retrospect. The only remaining digit becomes the associate of Y.

i.d.t. – ninetyfive

```
  S E V E N T E E N
  S E V E N T E E N
          T W E N T Y
              N I N E
              F I V E
              F I V E
              F I V E
              F I V E
                S I X
                S I X
-------------------
N I N E T Y F I V E
```

The leftmost column gives the value of N, followed by the determination of E from the ten millions column immediately thereafter. A unique selection for V emerges from observation of the millions column, and the hundred millions column yields the assignments for S and I without ambiguity. Go to the units column next to obtain the condition relating X and Y, noting the parity of Y from the same source. From the presence of Y in the ten thousands column, only two choices exist for this letter, each of which then generates a correspondent for X. Move to the tens column to secure the value of T in each instance, checking that the requirement of the hundreds column is met. From the thousands column, find the value of F in one case and eliminate the other case for lack of such a value. Lastly, note that the carryover into the ten thousands column is exactly the necessary one to associate the remaining digit with W.

dynamic duos

```
      W E' L L
  R E V E A L
        A L L
        T H E
-------------
  V O W E L S
```

The ten thousands column yields exact values for E and O. Then the value of V can be found from the thousands column, with the assignment for R consequently following from the leftmost column. Examination of the remaining digits reveals only one possible carryover into the hundreds column, thus establishing the value of the sum A + T. Choose the only pair of available digits that satisfies this relationship and impose the side condition to select the larger of the two for A. Pass to the units column to obtain two possible choices for the values of L and S, only one of which gives an appropriate value for H when the tens column is inspected. Finally, equate the only digit that remains to W.

i.d.t. – thousand

7(HUNDRED) + 20(THREE) + 24(TEN) = THOUSAND

Ten-shifts produce the equivalent problem listed below:

```
          E
  H U N D R   D
  H U N D R   D
  H U N D R   D
  H U N D R   D
  H U N D R   D
  H U N D R   D
  H U N D R E D
    T H R E E
    T H R E E
        T E N
        T E N
          T   N
          T   N
          T   N
          T   N
---------------
T H O U S A N D
```

Then the two rightmost columns offer values for the trio D, N, E, while the three columns on the left give possibilities for the trio H, T, U. Compare the two lists and find compatible choices from them. For each, determine the carryover into the hundred thousands column and use it together with the contents of that column to obtain O where available. Establish the condition relating R and A from the hundreds column and check the digit pool to see if it can be met. Wherever it can, proceed to the thousands column to verify if the value generated for S and the resulting carryover into the ten thousands column are appropriate. This turns out to be true once and only once.

marking time

```
    W A L T E R
        W I L L
      A L T E R
G I D G E T' S
D I G I T A L
```

From the hundred thousands column, find two possible candidates for the value of W. In both cases, the millions column establishes the relationship between G and D. Use it to determine the parity of the carryover into the ten thousands column, which in turn presents options for A. At this point, there are but three paths to pursue. In each instance, list available digit pairs for G and D according to the previously-noted equation. Then go to the thousands column to find viable assignments for L and consequently for I. Joint consideration of the tens and hundreds columns yields eligible values for E and T. In only one scenario will the two outstanding digits satisfy the requirement that the units column imposes upon R and S.

i.d.t. – ninety

```
T W E N T Y
T W E L V E
  S E V E N
    N I N E
    N I N E
    N I N E
    N I N E
    F I V E
    F I V E
    F I V E
N I N E T Y
```

Catalogue the permissible selections for E and N from the units column, observing that the list can be abbreviated due to the required parity of the carryover into the tens column. The value of N then implies the value of T upon examination of the leftmost column. Look at the tens column once again to find possible assignments for V. In each case, follow this by determining the relationship between L and I that the hundreds column suggests and satisfy it with unused digits whenever this can be accomplished. Continuing leftward, the thousands column fixes the value of F where available. The equation linking W and S that is obtained from the ten thousands column is solvable in a unique scenario. The argument culminates with the remaining digit being chosen as the value of Y.

i.d.t. – eightynine

```
  S E V E N T E E N
  S E V E N T E E N
    E I G H T E E N
          S E V E N
          S E V E N
          S E V E N
          S E V E N
              O N E
              O N E
              O N E
              O N E
              O N E
              O N E
              O N E
              O N E
              O N E
-------------------
E I G H T Y N I N E
```

Record the value of E from the leftmost column and use it to find N from the units column, noting that the carryover produced is the appropriate one to satisfy the requirement imposed by the tens column. Move to the hundred millions column to list options for the letters S and I. For each, examine the contents of the hundreds column to determine choices for O and V, and with the resulting carryover in mind, designate T from the thousands column where possible. The carryover into the millions column is now known. Find H from that column and follow this discovery by obtaining G from the ten millions column. Lastly, get the assignment for Y from the ten thousands column, observing that the correct carryover is generated to confirm the relationship demanded by the hundred thousands column.

in a word

```
L E S T E R
        I S
      T H E
      T R I
    G R A M
-----------
M A S T E R
```

Obtain the values of E and A from the ten thousands column, leaving only one possible choice for G when the thousands column is examined. Shift to the units column to establish the exact value of $S + I + M$ and thereby learn the carryover into the tens column. From that column, deduce that there are just two options for the value of $I + H + R$. List trios of available digits that would be suitable for these letters. Go next to the hundreds column to determine how R and T must be related. Noting the parity of R, find that but one of the trios offers a selection for R that yields an unassigned digit for T. Once R has been designated, the pair of digits associated with I and H becomes known. Look at the leftmost column to get the equation linking L and M, and observe at this pont that there is only one way to satisfy it. The sole remaining digit then becomes the correspondent of S. Finally, return to the units column to decide upon the value of I, from which the value of H follows.

fifty's not nifty

```
          I
  D R I V E
  T H E R E
        A T
        6 0
    T H E N
        A T
        4 0
          I
-----------
R E T U R N
```

The value of R, apparent from the hundred thousands column, leads to two possible choices for H when the thousands column is examined. In both cases, obtain the same equation involving D, T, and E from the ten thousands column. Keeping the choices for H in mind, establish a cap on the value of E. The constraint suggests starting with the largest possible value of E and working downward. For a specified value of E, list the options for the values of D and T from the aforementioned equation. In each case, explicitly note the two choices for H. Next, shift to the units column to find the value of I where available. The resulting carryover into the tens column and the contents of that column yield a relationship between V and A. Whenever the remaining digits permit its satisfaction, pass to the hundreds column to see if the needed value of U is still unassigned. Once this assignment is made, match the only digit left with N.

i.d.t. – hundred

$$10(\text{SEVEN}) + 4(\text{THREE}) + 2(\text{NINE}) = \text{HUNDRED}$$

A ten-shift converts the problem into

```
 S E V E N
   T H R E E
   T H R E E
   T H R E E
   T H R E E
     N I N E
     N I N E
-------------
H U N D R E D
```

and allows for the immediate discovery of H. The units column then produces candidates for E and D, with N following from the tens column as availability permits. In all cases, the actual carryover into the ten thousands column can be determined, but in only one will it be suitable for finding values of T from that column. For each such value, obtain explicit assignments for S and U from the hundred thousands column. Obtain the relationship between R and I from the hundreds column and observe that its satisfaction can be achieved in just one way in just one of the cases. The carryover into the thousands column establishes the lone remaining digit as the appropriate selection for V.

as a matter of fact

```
  Y O U R
  T U R N
      T O
    G E T
    T H E
---------
T R U T H
```

View the leftmost column first to find the value of T. The exact carryover into the thousands column can now be determined, thus leading to the assignments for R and Y. Careful scrutiny of the tens column reveals only one possibility for the value of the carryover into the hundreds column, and this yields the sum G + O explicitly. For each potential carryover from the units column, obtain the actual value of the sum U + E + H from the tens column. Then list pairs of available digits that satisfy the former, trios of available digits that satisfy the latter, and correlate these two lists. In each remaining case, write the equation involving the letters N, O, E, and H as suggested by the units column. Consideration of the options already noted for O, E, and H together with the constraint is enough to produce the unique situation in which the selection of N can be made.

rhyme time

```
  S I M P L E
          I T
          I S
  P O E T I C
-------------
L I C E N S E
```

The leftmost column immediately gives the value of L. Consideration of the carryover into the tens column and the contents of that column relate I and S. Observing that the hundred thousands column establishes the relative sizes of these letters, list the possibilities for their values. Also use the hundred thousands column to find acceptable values of P in each instance. Go next to the units column to secure the value of the sum T + C and appropriately select unused digits for these letters. There is but one case where available digits exist as assignments for M, N, and O when the hundreds, thousands, and ten thousands columns are checked, respectively. Finally, the single remaining digit is associated with E.

i.d.t. – ninetyeight

TWENTYTHREE + TWENTYNINE + 4(EIGHT) + 14(ONE)
= NINETYEIGHT

Use a ten-shift to rewrite the problem as

```
T W E N T Y T H R E E
  T W E N T Y N I N E
            E I G H T
            E I G H T
            E I G H T
            E I G H T
              O N E
                O N E
                O N E
                O N E
                O N E
---------------------
N I N E T Y E I G H T .
```

Then obtain the value of N from the millions column and follow this by finding the value of T from the leftmost column. Noting the required carryover into the hundred thousands column, go to the units column to determine E. Subsequently move left to discover the value of H from the tens column. The hundred millions column next establishes the value of W, which in turn gives the assignment for I when the billions column is considered. The remaining digits offer a unique selection for Y upon examination of the ten thousands column. Looking at the thousands column and the carryover into that column permits only one choice for O. Finally, the hundreds column yields an explicit equation that relates R and G which can be satisfied by the two digits that are still available.

identity crisis

```
    A B L E
  M A B E L
L A B E L S
      A L L
    2 5 0 0
-----------
C A B L E S
```

Joint inspection of the units and tens columns offers just two choices for the value of L, each of which yields one value of E from the units column alone. Only one of these options produces an acceptable assignment for C when the leftmost column is examined. Move to the right to first obtain the value of M from the ten thousands column, followed by the value of B from the hundreds column. Associate the one remaining digit with S.

generation gap

```
   T H I S
       I S
     W H Y
       I T
   I S N'T
----------
 F I S H Y
```

Note the value of F from the leftmost column and follow this by finding the value of T from the thousands column. The latter admits only one possible assignment for S when the units column is inspected. Then the contents of the tens column and the carryover into that column together establish a relationship between I and N. Keeping the constraint in mind, choose I accordingly and hence secure the value of N. Move next to the hundreds column to determine the value of the sum W + H and list available options for these digits. Once explicit designees have been selected for W and H, take the smallest remaining digit to be the value of Y.

all mixed up

```
  A N A G R A M S
  A N A G R A M S
  A N A G R A M S
  A N A G R A M S
  ---------------
  S C R A M B L E
```

The leftmost column offers two options for A. Pass to the ten thousands column to find possibilities for G, and for each, examine the hundred thousands column to determine R. Take this value to the thousands column to obtain the value of M where available, using the carryover into the ten thousands column to eliminate many of the cases that still exist at this point. Next, use the hundreds column to assign the value of B if possible. View the leftmost column once again to get the value of S. At this juncture, exactly one avenue remains open. The value of S generates the selection for E when the units column is considered. The tens column then yields the value of L, with the remaining pair of digits nicely fitting into the equation suggested by the millions column.

i.d.t. – ninetynine

```
  S E V E N T E E N
  S E V E N T E E N
  S E V E N T E E N
        F I F T E E N
            T H R E E
            T H R E E
            T H R E E
            T H R E E
            T H R E E
              N I N E
              N I N E
---------------------
N I N E T Y N I N E
```

The leftmost column in conjunction with both the units and tens columns secures the values of N and E. The relationship that results from the hundreds column then offers two possibilities for I, only one of which produces an acceptable assignment for S when the hundred millions column is inspected. The equation involving V and F that is inferred from the millions column can at this point be satisfied in a unique fashion. With the digits that remain available, use the hundred thousands column to determine the value of T. The carryover into the thousands column is established next, leading to the discovery of H from that same column. Once this is accomplished, go to the hundreds and ten thousands columns to find the correspondents for R and Y, respectively.

fourplay

```
 Y O U'L L
     U S E
   F O U R
 F O U R S
       O F
-----------
C O U R S E
```

Noting the value of C from the leftmost column, examine the thousands and ten thousands columns jointly to list options for the values of F and O. In each instance, find the value of Y from the ten thousands column and determine U when the carryover from the hundreds column is considered. The units column next yields possible values for the sum L + R + S. Moving to the left, the tens column gives the value for L + R. Simultaneous solution of these two equations establishes the value of S. Once this is known, assign digits to L and R where available to meet the needs of L + R. In only one case will the hundreds column produce an appropriate choice for R, with the value of L then following. At this point, two digits remain. Select E to accommodate the initial restriction.

i.d.t. – sixtysix

2(NINETEEN) + 5(THREE) + SIX + 7(ONE) = SIXTYSIX

After a ten-shift, the problem looks like this:

```
            E
N I N E T E E N
N I N E T E E N
      T H R E
      T H R E
      T H R E
      T H R E
      T H R E
          S I X
          O N
          O N
          O N
          O N
          O N
          O N E
          O N E
---------------
S I X T Y S I X
```

The units column permits three choices for the sum N + E, one of which is immediately eliminated due to the presence of N in the leftmost column. In each case that remains, the value of N + E together with the carryover into the tens column will yield E from that column. This gives N in retrospect, from which the millions and leftmost columns produce assignments for I and S, respectively. Next, relate R and O from the hundreds column and note that the parity of R can be determined from it. This turns out to be sufficient to find O explicitly. A joint consideration of the tens thousands and hundred thousands columns reveals the values of T and X. Then return to the hundreds column to list the options for R. From the thousands column, observe that the two digits still available can be suitably associated with H and Y in precisely one of the existing situations.

i.d.t. – ninetytwo

$$FORTYNINE + TWENTYONE + 4(THREE) + TWO + 8(ONE) = NINETYTWO$$

Two ten-shifts recast the problem as

```
F O R T Y N I
T W E N T Y O
        T H R E E
        T H R E E
        T H R E E
        T H R E E
            T W O
            N E
            O
            O
            O
            O
            O
            O
            O
            O
-----------------
N I N E T Y T W O .
```

Then note the value of E from the units and tens columns. The sum of T and N is restricted by the hundred thousands column and an order relationship involving these letters follows from the leftmost column. List all options and determine R from the millions column where possible. Obtain choices for the assignment for F from the leftmost column as availability permits. In each instance, establish the condition imposed upon O and I by the hundreds column. Observing their relative sizes from the ten millions column, find viable selections for these letters where this can be accomplished. In exactly one case will the ten millions column yield W, the thousands column, H, and the ten thousands column, Y, in an appropriate fashion.

whatchamacallits

```
 T H A T' S
         I T
         I S
     T H A T
     W H A T
       I T' S
 -----------
C A L L E D
```

The values of C and A are discernible at the start from the two leftmost columns, with the ten thousands column offering but two options for T. In each case, the units column establishes a connection between S and D. List possible alternatives that satisfy it and follow this by examining the tens column, thereby linking I and E. Whenever assignments for this pair can be made from the remaining pool of digits, go into the hundreds column to relate H and L and designate values for these letters accordingly. The contents of the thousands column, together with the carryovers into and from it, will produce an acceptable choice for W in exactly one instance.

home sweet home

```
W H I C H
    O N E
  O W N S
    T H E
W H I T E
---------
H O U S E
```

Begin by obtaining the values of W and H from consideration of the ten thousands column and the carryover into that column. The choice for H then dictates the carryover from the hundreds column, thus implying a unique assignment for I. Furthermore, the hundreds column restricts the possible values for the pair O and T. Establish the relationship linking S and E from the units column and in each case, satisfy it with the digits still available. Fixing the value of T (and thus the value of O), move into the tens column to find appropriate values for C and N. In exactly one instance will the carryover into the hundreds column properly yield the last remaining digit as the value of U.

i.d.t. – ninetysix

2(NINETEEN) + EIGHTEEN + 10(THREE) + TEN = NINETYSIX

Employ a single ten-shift to rewrite the problem as

```
  N I N E T E E N
  N I N E T E E N
  E I G H T E E N
      T H R E E
            T E N
-----------------
N I N E T Y S I X
```

and immediately note the values of N and X from the leftmost and units column, respectively. The tens column allows two choices for I, only one of which is realistic in light of the millions column. The ten millions column then establishes the digital correspondent of E. A review of the hundreds column next relates T and S. Make assignments for these letters according to availability. The value of T gives the parity of the carryover into the ten thousands column and thereby secures the actual carryover. Knowledge of this fact produces the value of H from that column, which subsequently yields G when the hundred thousands column is investigated. In each of the existing cases, observe the relationship between R and Y that the thousands column demands, and find satisfaction of it in exactly one instance.

ready, set, go

```
    N A M E
      A L L
  E I G H T
          N
-----------
S T A T E S
```

The values of S, E, and T are immediately apparent from the ten thousands and hundred thousands columns. Go next to the units column and use the preassigned value of N to disclose the value of L. Then the tens column leads to only one possibility for the sum M + H, which can be achieved by only one pair of available and interchangeable digits. Pass to the hundreds column to determine the digits that correspond to A and G and subsequently obtain the value of I from the thousands column.

i.d.t. – eightytwo

3(THIRTEEN) + 9(THREE) + TEN + 3(TWO) = EIGHTYTWO

Employ a ten-shift to transform the problem into

```
            E
  T H I R T E   N
  T H I R T E   N
  T H I R T E   N
        T H R   E
        T H R   E
        T H R   E
        T H R   E
        T H R   E
        T H R   E
        T H R   E
        T H R E E
        T H R E E
            T E N
            T W O
            T W O
            T W O
-----------------
E I G H T Y T W O .
```

The parity of E is predictable from the units column. Coupling this with the presence of E in the leftmost column is enough to get its value explicitly. Relate N and O from the units column and note the requisite parity of the carryover into the tens column. Each choice of digits for these letters leads to the determination of values of W upon examination of the tens column. Proceed leftward to the hundreds column to obtain the condition involving R and T. From the ten millions column, establish a minimal value of T, thus limiting the options for the actual assignments of T and R. The same source produces values for I as availability allows. Then use the carryover into the hundred thousands column to secure H and follow this by finding G from the millions column where possible. Complete this argument by observing that the thousands column yields the correct value of Y as well as the appropriate carryover into the ten thousands column in just one case.

nothing but the truth

```
      I  S
      I  T
M A G I  C
      N  O
    I T' S
----------
L O G I  C
```

Learn the values of A and O from the thousands column. Then move leftward to establish the equation relating M and L. Examination of the hundreds column and the carryover into it offers only one option for the value of I. As a result, the tens column gives the value of the sum N + T, which can be achieved with a unique pair of unassigned digits. Find the parity of T from the units column and thereby get specific values of T and N. A second glance at the units column now yields the value of S. At this point, the aforementioned equation involving M and L can be satisfied in two ways. Select the one that conforms to the constraint, and use the constraint once more to appropriately choose the values of G and C.

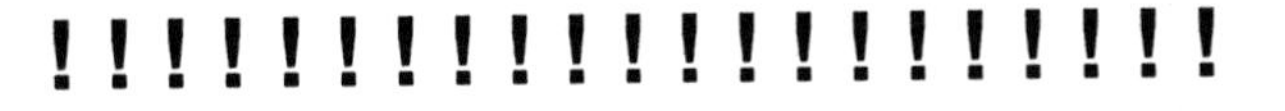

SECTION 3

SOLUTIONS

! !

in a word

```
  3 9 6 7 9 5
          1 6
        7 2 9
        7 5 1
      8 5 0 4
  -----------
  4 0 6 7 9 5
```

1. unBEKnowst
2. coCCYx
3. joDHPurs
4. quEUE
5. aFGHan
6. sovereiGNTy
7. cHIHuahua
8. sheIKDom
9. radIOActive
10. dooMSDay
11. viNTNer
12. boOKKeeper
13. brOUHaha
14. driVEWay
15. baZAAr
16. schiZOPhrenia

(Answers are not necessarily unique.)

i.d.t. – ninetytwo

```
  6 9 7 1 2 8 3 8 0
  1 4 0 8 1 2 9 8 0
          1 5 7 0 0
          1 5 7 0 0
          1 5 7 0 0
          1 5 7 0 0
              1 4 9
              9 8 0
              9 8 0
              9 8 0
              9 8 0
              9 8 0
              9 8 0
              9 8 0
              9 8 0
  -----------------
  8 3 8 0 1 2 1 4 9
```

i.d.t. – sixtynine

```
  1 2 3 2 4 7 2 2 4
    4 6 4 2 7 2 2 4
          7 8 0 2 2
              7 2 4
              7 2 4
              7 2 4
  -----------------
  1 6 9 7 5 4 6 4 2
```

on a count

```
    8 0 8 5 2
    8 9 7 4 7
      1 7 2 4
        5 3 1
  -----------
  1 7 2 8 5 4
```

(0 and 9 interchangeable)

There are 170 triangles shown – 50 dels and 120 deltas.

BONUS: The total number of triangles is given by

$$\frac{2n^3 + 5n^2 + 2n - \mu}{8}$$

$$\text{where } \mu = \begin{cases} 1 \text{ if n is odd} \\ 0 \text{ if n is even.} \end{cases}$$

identity crisis

```
        3 4 6 8
      9 3 4 8 6
    6 3 4 8 6 1
          3 6 6
        2 5 0 0
  -------------
    7 3 4 6 8 1
```

To be definite, assume that the labels "3B," "3W," and "2B,1W" appear on cartons #1, #2, and #3, respectively. Open carton #3 and withdraw one cable from it. Since the label on the carton is incorrect, the selected cable consists of either three black wires or three white wires.

If the former is true, then carton #3 should have the label "3B." Since carton #2 is also mismarked, it should wear the label "2B,1W," leaving the "3W" label for carton #1.

If the latter is true, then the correct label for carton #3 would be "3W." Hence, carton #1, whose current label is incorrect, should be marked "2B,1W," implying that "3B" is the proper label for carton #2.

i.d.t. – ninetyeight

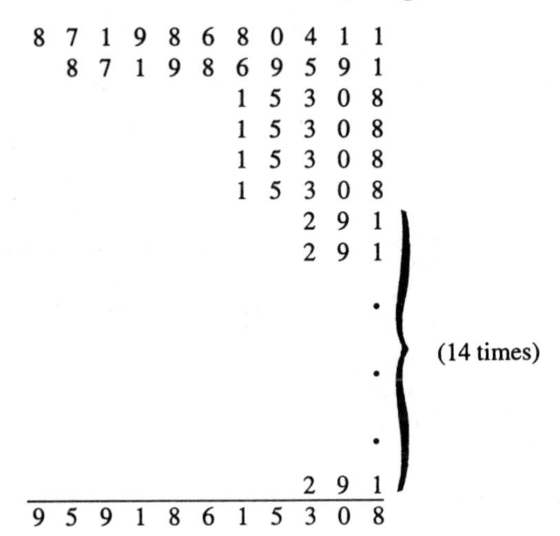

dynamic duos

```
     3 9 2 2
 7 9 8 9 6 2
       6 2 2
       4 1 9
 -----------
 8 0 3 9 2 5
```

1. ROMEO and JULIET
2. LAUREL and HARDY
3. SPAIN and PORTUGAL
4. GILBERT and SULLIVAN
5. SALT and PEPPER
6. LADIES and GENTLEMEN
7. REST and RELAXATION
8. BULLS and BEARS
9. MUSTARD and SAUERKRAUT
10. STARS and STRIPES
11. DEATH and TAXES
12. PRIDE and PREJUDICE
13. THUNDER and LIGHTNING
14. LEWIS and CLARK
15. PINS and NEEDLES
16. CHURCH and STATE
17. MEATBALLS and SPAGHETTI
18. CLOAK and DAGGER
19. STICKS and STONES
20. TWEEDLEDEE and TWEEDLEDUM

i.d.t. – eightytwo

```
  9 4 8 7 9 2 2 0
  9 4 8 7 9 2 2 0
  9 4 8 7 9 2 2 0
        9 4 7 2 2
        9 4 7 2 2
        9 4 7 2 2
        9 4 7 2 2
        9 4 7 2 2
        9 4 7 2 2
        9 4 7 2 2
        9 4 7 2 2
        9 4 7 2 2
            9 2 0
            9 6 1
            9 6 1
            9 6 1
-----------------
2 8 5 4 9 3 9 6 1
```

preface

```
3 6 2 5 0
  7 8 5 3
---------
4 5 2 1 3
```

(6 and 7 interchangeable)

i.d.t. – eightynine

```
 5 1 2 1 6 8 1 1 6
 5 1 2 1 6 8 1 1 2
   1 0 3 4 8 1 1 6
         5 1 2 1 6
         5 1 2 1 6
         5 1 2 1 6
         5 1 2 1 6
             7 6 1
             7 6 1
             7 6 1
             7 6 1
             7 6 1
             7 6 1
             7 6 1
             7 6 1
             7 6 1
-------------------
1 0 3 4 8 9 6 0 6 1
```

from a to z

```
   5 3 4
 1 0 2 5
 4 6 7 7
   3 0 8
 4 3 8 1
---------
1 0 9 2 5
```

(7 and 8 interchangeable)

The worm munches the front covers of Volumes 1–9, the back covers of Volumes 2–10, and the pages of Volumes 2–9. This means that the entire horizontal path measures $9\left(\frac{1}{9}\right)+9\left(\frac{1}{9}\right)+8\left(\frac{7}{2}\right) =$ $1 + 1 + 28 = 30$ inches.

i.d.t. – eightysix

```
7 8 3 8 6 9 8 8 6
    7 0 2 9 8 8 6
    7 0 2 9 8 8 6
    7 0 2 9 8 8 6
        7 8 3 8 6
        7 8 3 8 6
        7 8 3 8 6
-----------------
8 0 5 1 9 4 7 0 2
```

fifty's not nifty

```
          6
  8 1 6 9 4
  5 7 4 1 4
        2 5
        6 0
    5 7 4 3
        2 5
        4 0
          6
-----------
1 4 5 0 1 3
```

Let x = average speed for the round-trip and let D = distance between towns A and B. Then, the time spent driving from A to B is $\frac{D}{60}$, while the time spent driving from B to A is $\frac{D}{40}$. Since the round-trip distance is D + D = 2D and the time spent driving the round-trip is $\frac{D}{60}+\frac{D}{40}=\frac{5D}{120}=\frac{D}{24}$, it follows that $x = \frac{2D}{\left(\frac{D}{24}\right)} = 48$ mph.

fourplay

```
  5 4 9 2 2
      9 7 6
    8 4 9 3
  8 4 9 3 7
        4 8
-----------
1 4 9 3 7 6
```

$$89 = 4! + (4! + \sqrt{4}) \div .4$$
$$104 = (4 \div .4)^{\sqrt{4}} + 4$$
$$109 = (44 - .4) \div .4$$
$$127 = (4^4 - \sqrt{4}) \div \sqrt{4}$$
$$142 = (4 + \sqrt{4})(4!) - \sqrt{4}$$
$$150 = (\sqrt{4} \div .4)! \div (.4 + .4)$$

These results are not necessarily unique.

generation gap

$$\begin{array}{r} 9\,0\,4\,6 \\ 4\,6 \\ 8\,0\,2 \\ 4\,9 \\ 4\,6\,5\,9 \\ \hline 1\,4\,6\,0\,2 \end{array}$$

Suppose x = number of fish that Jack caught
y = number of fish that Jack's father caught
z = number of fish that John caught
w = number of fish that John's father caught,

where x, y, z, w are nonnegative integers. Then $y + w = 2(x + z)$, $y = 2w$, and $x + y + z + w = 35$. From the first and third equations, it follows that $3(x + z) = 35$, which cannot be satisfied by nonnegative integers. Conclude that there are not four people on the fishing trip.

Another possibility is that Jack and John are brothers and have the same father. However, the problem refers to "the fathers," so this option can be eliminated.

Now suppose that John is Jack's father.

Let x = number of fish that Jack caught
y = number of fish that Jack's father (John) caught
z = number of fish that John's father caught,

where x, y, and z are nonnegative integers. Then $y + z = 2(x + y)$, $y = 2z$, and $x + y + z = 35$. The first of these yields $z = 2x + y$. Substituting $2z$ for y in this equation gives $2x + z = 0$, which can only be true if $x = z = 0$. Then $y = 2(0) = 0$ as well, leading to a contradiction of the third equation in the system.

The only case left to consider is the one that assumes that Jack is John's father. Accordingly,

Let x = number of fish that Jack (John's father) caught
y = number of fish that Jack's father caught
z = number of fish that John caught.

Then $x + y = 2(x + z)$, $y = 2x$, and $x + y + z = 35$, from which it follows that $x = 10$, $y = 20$, and $z = 5$. Since this is the only solution found, Jack is John's father. Therefore, Jack is older.

i.d.t. – fourhundredfortyone

```
9 7 5 6 8 5 0 4 6 1 4 9 7 5 6 3 1 1 0
                            3 8 6 1 1
                              9 7 5 6
                              9 7 5 6
                              9 7 5 6
                                3 1 0
                                7 0 1
                                7 0 1
-------------------------------------
9 7 5 6 8 5 0 4 6 1 4 9 7 6 3 2 7 0 1
```

as a matter of fact

```
  8 5 4 0
  1 4 0 7
      1 5
    2 6 1
    1 9 6
---------
1 0 4 1 9
```

Rotate the page 180 degrees to obtain

X = I + IX

whatchamacallits

```
  9 3 0 9 7
        6 9
        6 7
    9 3 0 9
    2 3 0 9
      6 9 7
-----------
1 0 5 5 4 8
```

1-j, 2-d, 3-g, 4-m, 5-h, 6-k, 7-a, 8-o, 9-1, 10-e, 11-n, 12-c, 13-i, 14-b, 15-f

dedication

```
  9 2 4 9 0
  1 7 2 6 6
-----------
1 0 9 7 5 6
```

i.d.t. – ninetyone

```
8 1 7 9 8 0 8 1 4
  6 4 3 2 8 7 7 9
  6 4 3 2 8 7 7 9
    6 5 6 8 7 7 9
    6 5 6 8 7 7 9
            8 1 4
            8 1 4
            8 1 4
            8 1 4
            8 1 4
            4 9 7
-----------------
9 5 9 7 8 0 4 9 7
```

i.d.t. – ninetynine

```
  6 7 4 7 2 5 7 7 2
  6 7 4 7 2 5 7 7 2
  6 7 4 7 2 5 7 7 2
      3 0 3 5 2 2 7
          5 8 9 7 7
          5 8 9 7 7
          5 8 9 7 7
          5 8 9 7 7
          5 8 9 7 7
              2 0 2 7
              2 0 2 7
-------------------
2 0 2 7 5 1 2 0 2 7
```

ready, set, go

```
   8 2 6 9
     2 4 4
 9 3 5 7 0
         8
-----------
1 0 2 0 9 1
```

(6 and 7 interchangeable)

1. { Athos, Porthos, Aramis }
2. { Soprano, Tenor, Alto, Bass }
3. { Huron, Ontario, Michigan, Erie, Superior }
4. { Lincoln, Jefferson, F. Roosevelt, Washington, Kennedy, Eisenhower }
5. { Dopey, Sleepy, Doc, Grumpy, Happy, Sneezy, Bashful }
6. { Nebraska, Nevada, New Hampshire, New Jersey, New Mexico, New York, North Carolina, North Dakota }
7. { Mercury, Venus, Earth, Mars, Jupiter, Saturn, Uranus, Neptune, Pluto }
8. { Alberta, British Columbia, Manitoba, New Brunswick, Newfoundland, Nova Scotia, Ontario, Prince Edward Island, Quebec, Saskatchewan }
9. { Alabama, Arkansas, Florida, Georgia, Louisiana, Mississippi, North Carolina, South Carolina, Tennessee, Texas, Virginia }
10. { Aries, Taurus, Gemini, Cancer, Leo, Virgo, Libra, Scorpio, Sagittarius, Capricorn, Aquarius, Pisces }
11. { Connecticut, Delaware, Georgia, Maryland, Massachusetts, New Hampshire, New Jersey, New York, North Carolina, Pennsylvania, Rhode Island, South Carolina, Virginia }
12. { Atlanta Braves, Chicago Cubs, Cincinnati Reds, Colorado Rockies, Florida Marlins, Houston Astros, Los Angeles Dodgers, Montreal Expos, New York Mets, Philadelphia Phillies, Pittsburgh Pirates, San Diego Padres, San Francisco Giants, St. Louis Cardinals }

i.d.t. – seventy

```
  2 0 6 8 2 5
  6 4 6 7 6 8
  6 4 6 7 6 8
    2 9 3 6 6
    2 9 3 6 6
    2 9 3 6 6
    2 9 3 6 6
    2 9 3 6 6
    2 9 3 6 6
        2 6 8
-------------
1 6 7 6 8 2 5
```

rhyme time

```
  9 2 0 3 1 7
          2 5
          2 9
  3 4 7 5 2 6
-------------
1 2 6 7 8 9 7
```

1. flabby tabby
2. fender bender
3. funny money
4. legal eagle
5. local yokel
6. creature feature
7. million billion
8. bargain jargon
9. ocean motion
10. fervent servant
11. Mexican lexicon
12. Caracas maracas

greetings

```
    2 5 1
2 5 0 6 6
2 5 0 7 1
        4
---------
5 0 3 9 2
```

Chan used the following argument to deduce that each married couple participated in a total of eight handshakes.

The individual who took part in eight handshakes necessarily shook the hand of each guest excepting his or her spouse. Consequently, only his or her spouse could have turned in the slip with "0" written on it, since everyone else would have been involved in at least one handshake. This means that the slips bearing the digits "0" and "8" were submitted by one of the other four married couples at the party. (Remember that Chan's slip is not part of the collection under scrutiny.)

Focus next on the person who participated in seven handshakes. Excluding his or her spouse as well as the couple who submitted the slips marked "0" and "8," this individual exchanged greetings with everyone else in the group. Since each of the six people that comprise this subset has now participated in at least two handshakes (with "7" and "8"), it follows that only the spouse of the person who wrote "7" could possibly have written "1." That is, the slips that contain the digits "1" and "7" had to have been handed in by another of the four married couples.

Continuing along this logical road, we see that slips "2" and "6" were submitted by a married couple, as were slips "3" and "5." This means that the only member of the original set of nine whose spouse is not in that set, namely Mrs. Chan, must have written the digit "4" on her slip.

i.d.t. – eighty

```
  7 0 6 1 }
  7 0 6 1 }
        . }  (12 times)
        . }
        . }
  7 0 6 1 }
    9 8 1 }
    9 8 1 }
        . }  (20 times)
        . }
        . }
    9 8 1 }
-----------
1 0 4 3 5 2
```

marking time

```
  9 3 8 0 7 4
      9 1 8 8
    3 8 0 7 4
5 1 6 5 7 0 2
-------------
6 1 5 1 0 3 8
```

The current time is 1:33 P.M. When a "3" is added to the far right of this display, the new reading becomes 13:33, which is 1:33 P.M. in military time.

i.d.t. – fiftytwo

```
2 7 4 5 2 9 9 6
  3 4 3 2 9 9 6
  3 4 3 2 9 9 6
          2 8 1
          2 8 1
          2 8 1
          2 8 1
          1 6 9
---------------
3 4 3 2 0 2 8 1
```

cover

```
3 8 9 5
      8
1 0 0 9
    8 3
      8
2 0 0 6
-------
7 0 0 9
```

(1 and 2 interchangeable)
(5 and 6 interchangeable)

array, array

```
    6 0 5
  9 0 3 7
    7 0 5
---------
1 0 3 4 7
```

The six numbers that Vic should have chosen are 6, 8, 17, 25, 30, and 49.

i.d.t. – ninety

```
2 4 3 6 2 9
2 4 3 0 5 3
  8 3 5 3 6
    6 1 6 3
    6 1 6 3
    6 1 6 3
    6 1 6 3
    7 1 5 3
    7 1 5 3
    7 1 5 3
-----------
6 1 6 3 2 9
```

nothing but the truth

```
      8 4
      8 2
6 9 5 8 3
      1 0
    8 2 4
---------
7 0 5 8 3
```

If one and only one statement is true, then exactly 8 – 1 = 7 of the statements must be false. Therefore, statement (7) is the true statement.

pun, pun, pun

```
  7 8 9 7
2 5 0 7 6
  1 6 5 3
---------
3 4 6 2 6
```

The ranch should be named "The Horizon" because that is the place where the sun's rays meet.

i.d.t. – hundred

```
    8 7 3 7 5
    8 7 3 7 5
    8 7 3 7 5
    8 7 3 7 5
    8 7 3 7 5
    8 7 3 7 5
    8 7 3 7 5
    8 7 3 7 5
    8 7 3 7 5
    8 7 3 7 5
    4 1 9 7 7
    4 1 9 7 7
    4 1 9 7 7
    4 1 9 7 7
      5 6 5 7
      5 6 5 7
-------------
1 0 5 2 9 7 2
```

i.d.t. – sixtysix

```
1 0 1 4 7 4 4 1
1 0 1 4 7 4 4 1
      7 5 3 4 4
      7 5 3 4 4
      7 5 3 4 4
      7 5 3 4 4
      7 5 3 4 4
          2 0 6
          9 1 4
          9 1 4
          9 1 4
          9 1 4
          9 1 4
          9 1 4
          9 1 4
---------------
2 0 6 7 8 2 0 6
```

home sweet home

```
2 5 0 4 5
    1 6 9
  1 2 6 7
    3 5 9
2 5 0 3 9
---------
5 1 8 7 9
```

The chart below is consistent with all of the given data.

	#1	#2	#3	#4	#5
owner	American	Frenchman	Englishman	Russian	Canadian
color	yellow	blue	red	white	green
pet	parrot	horse	cat	dog	turtle
car	Buick	Ford	Cadillac	Chevrolet	Pontiac
liquor	vodka	scotch	bourbon	gin	rum

Accordingly, the Canadian owns the turtle, the American drinks vodka, and the Russian owns the white house.

i.d.t. – thousand

```
     8 1 4 9 0 3 9
     8 1 4 9 0 3 9
     8 1 4 9 0 3 9
     8 1 4 9 0 3 9
     8 1 4 9 0 3 9
     8 1 4 9 0 3 9
     8 1 4 9 0 3 9
         5 8 0 3 3 ⎫
         5 8 0 3 3 ⎪
                 . ⎬ (20 times)
                 . ⎪
                 . ⎪
         5 8 0 3 3 ⎭
             5 3 4 ⎫
             5 3 4 ⎪
                 . ⎬ (24 times)
                 . ⎪
                 . ⎪
             5 3 4 ⎭
   ---------------
   5 8 2 1 6 7 4 9
```

i.d.t. – ninetysix

```
   1 0 1 8 6 8 8 1
   1 0 1 8 6 8 8 1
   8 0 7 9 6 8 8 1
         6 9 3 8 8
         6 9 3 8 8
         6 9 3 8 8
         6 9 3 8 8
         6 9 3 8 8
         6 9 3 8 8
         6 9 3 8 8
         6 9 3 8 8
         6 9 3 8 8
         6 9 3 8 8
             6 8 1
-----------------
1 0 1 8 6 5 2 0 4
```

threesomes

```
  7 2 1 8 3 0 9
  7 2 1 8 3 0 9
  7 2 1 8 3 0 9
---------------
2 1 6 5 4 9 2 7
```

The woman gives birth aboard a cruise ship. The first of the babies is born at 11:57 P.M. on a Monday. One minute later, the ship crosses the International Date Line. The second baby is born at 11:59 P.M. on what is now a Tuesday. The last of the births occurs two minutes later, at 12:01 A.M. on Wednesday.

all mixed up

```
2 4 2 8 1 2 5 9
2 4 2 8 1 2 5 9
2 4 2 8 1 2 5 9
2 4 2 8 1 2 5 9
---------------
9 7 1 2 5 0 3 6
```

The letters of RODE NOW can be rearranged to form ONE WORD.

i.d.t. – ninetyfive

```
  6 5 7 5 1 9 5 5 1
  6 5 7 5 1 9 5 5 1
        9 2 5 1 9 8
            1 3 1 5
            4 3 7 5
            4 3 7 5
            4 3 7 5
            4 3 7 5
              6 3 0
              6 3 0
-------------------
1 3 1 5 9 8 4 3 7 5
```

solutions chart

The following symbols are used in the chart:

* each solution has an interchangeable pair of digits
** each solution has two interchangeable pairs of digits

alphametic	number of solutions
cover	one**
dedication	five
preface	two*

narrative

alphametic	number of solutions
greetings	four
threesomes	two
from a to z	six*
pun, pun, pun	twelve*
on a count	four*
array, array	four
dynamic duos	two
marking time	one
in a word	one
fifty's not nifty	three
as a matter of fact	two
rhyme time	one
identity crisis	four
generation gap	fifteen
all mixed up	one
fourplay	two
whatchamacallits	one
home sweet home	one
ready, set, go	two*
nothing but the truth	four

alphametic	number of solutions
ideal doubly-true	
i.d.t. – fiftytwo	one
i.d.t. – eightysix	one
i.d.t. – fourhundredfortyone	one
i.d.t. – sixtynine	one
i.d.t. – eighty	one
i.d.t. – seventy	one
i.d.t. – ninetyone	one
i.d.t. – ninetyfive	one
i.d.t. – thousand	one
i.d.t. – ninety	one
i.d.t. – eightynine	one
i.d.t. – hundred	one
i.d.t. – ninetyeight	one
i.d.t. – ninetynine	one
i.d.t. – sixtysix	one
i.d.t. – ninetytwo	one
i.d.t. – ninetysix	one
i.d.t. – eightytwo	one

about the author

When the telephone company switched its exchange codes from letters to numbers in the 1960s, Steven Kahan was motivated to do the same with ordinary addition examples. He's been doing so ever since, creating several thousand alphametics in the process while serving as the Alphametics Editor of the *Journal of Recreational Mathematics*. To finance his avocation, he has been teaching at Queens College of the City University of New York for the past quarter century, where he tries to inject the love of mathematics into the fertile minds of his students.

He shares his living space with his wife, Susan, and his two children, David and Sara. He is currently compiling a massive compendium of all ideal, doubly-true alphametics known to man and computer.